THE BOOK OF
FORCES

METIN BEKTAS

DEDICATION

This book is dedicated to all curious people.

CONTENTS

Part One

Driving for hours at constant speed on a highway can be pretty boring. The same is true for airplane travel. While the take-off and landing phase is most certainly thrilling, once you reach the cruising altitude the butterfly sensation is gone. Shouldn't traveling at velocities close to the speed of sound be more exciting? Excitement is usually associated with high speeds, but in reality, it is never the speed that gives us the exhilarating sensation of extreme motion. We simply don't have the ability to feel speed. You could be traveling at velocities close to the speed of light and you wouldn't feel a thing. What we are able to feel however is acceleration, the change in speed.

When a roller coaster train reaches and passes the highest point, the gravitational force causes it to gain speed. The train then goes from rest (0 meters per second or in short: 0 m/s) to the highway speed limit (75 mph $\approx$ 33 m/s) in a second or less. Our inner ear picks up this rapid change in speed and sends this information to the brain. The result: pure, uncut excitement that more than makes up for the hour we spent waiting in line. So acceleration, the rate of change in speed, is not just an abstract mathematical quantity used by scientists in tedious calculations, it is a fundamental part of the human experience.

Computing acceleration is not that difficult. You simply divide the change in velocity (symbolized by Δv and measured in m/s) by the elapsed time (symbolized by Δt and measured in s). The resulting unit is meters per square second, m/s². Denoting the acceleration by the letter a, we can also write this definition in form of a simple equation:

$$a = \Delta v / \Delta t$$

A value of 5 m/s² for example means that the velocity goes up by 5 m/s over the course of one second. So an object accelerating from rest with a constant acceleration of 5 m/s² will have the speed 5 m/s (11 mph) after one second, 10 m/s (23 mph) after two seconds, 15 m/s (34 mph) after three seconds, and so on. As long as the motor or engine manages to maintain this level of acceleration, the velocity will keep increasing in this fashion.

A very important value of acceleration you should always keep in mind is 10 m/s². Why? This is the acceleration experienced by an object in free fall at the surface of Earth when air resistance is negligible. For example, if you drop a coin from a tower or a tall building, it will have a speed of 10 m/s (23 mph) after one second, 20 m/s (45 mph) after two seconds, 30 m/s (68 mph) after three seconds, and so on. Since we roughly know what this level of acceleration feels like, from jumping down a chair or falling down the stairs (which is considerably less pleasant), it makes for a great reference value. So instead of using the rather abstract unit meters per square second, you can also express acceleration in g's, with 1 g being equal to 10 m/s². However, keep in mind that when inputting a value for the acceleration into a formula, you always need return to the basic unit m/s².

Let's go back to the roller coaster to see how we can use the above definition to calculate acceleration. Suppose that the roller coaster goes from rest to 33 m/s in just 0.7 seconds. What is the corresponding acceleration? Since the change in velocity is Δv = 33 m/s and the elapsed time Δt = 0.7 s, we get a = Δv / Δt = 33 / 0.7 m/s^2 ≈ 47 m/s^2 or 4.7 g's - even more exciting than free fall. However, such high levels of acceleration can only be endured for a short time. If maintained for several seconds or more, loss of consciousness is inevitable. Such acceleration-induced blackouts are especially dangerous for fighter pilots and astronauts, which is why learning to tolerate rapid changes in speed is a vital part of their training.

Note that acceleration can also take on negative values, in which case it is often referred to as deceleration. The change in speed Δv is always calculated by subtracting the initial velocity from the final velocity (change equals final minus initial). If the final speed is lower than the initial speed, the change in speed, and thus the acceleration, will be negative. Also note that the special case of zero acceleration implies motion with constant velocity.

Acceleration goes beyond pure changes in speed. When a body travels along a curved path (as opposed to a straight line), it will be subject to acceleration even if its velocity remains constant. The magnitude of the resulting acceleration depends on the velocity v (in m/s) and the radius of curvature r (in m). A higher velocity and smaller radius lead to a greater acceleration. The corresponding formula is:

a = v^2 / r

In words: we square the velocity and divide that by the radius of curvature (which is generally variable along a path, but constant in case of motion on a circle) - quite simple. For example, traveling at a constant speed of $v = 30$ m/s through a curve of radius $r = 300$ m implies an acceleration of $a = v^2 / r = 30^2 / 300$ m/s^2 = 3 m/s^2 = 0.3 g's. For the most part of the book we will stick to the "one-dimensional" definition of acceleration being equal to the rate of change in speed with respect to time, but the fact that curved motion always implies acceleration is worth keeping in mind.

Though not vital for our discussion of force, this seems like the appropriate time to mention yet another interesting property of acceleration. You've probably heard the statement "speed is relative" many times. How fast are you traveling at this moment? Well, you're sitting on your couch reading this book, so it seems that you are at rest. An observer located on the Sun would naturally disagree. She would see you rotating about Earth's axis and moving in an orbit around the Sun at the same time. According to her, you are swooshing through space at mind-blowing velocities. And an observer in the center of our galaxy would assign you an even higher speed as our solar system is also on the move when viewed from said center. Who is right? Are you moving or are you at rest? The answer depends on your point of reference and without one, the quantity speed is absolutely meaningless.

What about acceleration? Suppose you are in a train with a cup of coffee. When the train is moving at a constant velocity, the surface of the coffee will be horizontal. It is impossible to deduce from the surface what your current speed is as the liquid surface will be horizontal for all speeds (assuming the speed is kept at the same level over time). However, when the train accelerates, the surface of the coffee will make an angle to the horizontal. You can measure this angle and compute the acceleration. Now suppose an observer on the Sun were able to see you and your coffee cup using an enormously powerful telescope. She would see the liquid surface at the same angle to the horizontal and deduce the same acceleration. Hence, the value of the acceleration someone or something experiences does not depend on the point of reference.

Why all this talk about acceleration? Isn't this book supposed to be about force? Indeed it is. But acceleration is the number one physical quantity you should be familiar with in order to understand what force is. So it pays to have a clear picture of what acceleration means (which I hope you now have) and how to calculate it before moving on to force.

Newton's Second Law

Suppose you have two identical cars and you equip them with different motors, one motor producing a weak force, the other a strong force. Clearly the car that is subject to the stronger force will experience a greater acceleration. More force means more acceleration - fair enough. But there's more. Suppose now instead of two identical cars you are given two identical motors (producing the same force) and you put one of the motors in a car and the other in a truck. Despite the same force input, the acccleration will obviously not be same. Mass acts as resistance to changes in motion and as a result of that, the acceleration of the truck is smaller. So force is intimately connected to both acceleration and mass.

Newton recognized this and provided the world with a neat definition of force. Newton's second law, we'll get to first law in a moment, states that if the mass of a body remains constant during motion, force F (in Newtons = N) is just the product of acceleration a (in m/s²) and mass m (in kg):

$F = m \cdot a$

So if we see a m = 1000 kg car that accelerates at a = 2 m/s², we can conclude that the net force acting on the car must be F = 1000 kg · 2 m/s² = 2000 N. The term net force refers to the sum and difference of all forces acting on a body. A 2000 N net force might be the result of just one 2000 N force pushing the body or it might the result of a 12000 N force pushing the body and a 10000 N force pulling on it at the same time. What counts in terms of motion is the overall force.

11

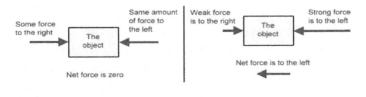

(Concept of net force)

This immediately leads to Newton's First Law: if the net force acting on a body is zero, meaning that either there's no force present or the forces pushing the body are equal to the forces pulling on it (force equilibrium), the resulting acceleration is zero and the body thus moves at constant speed. For example, when a car travels at constant speed, the force provided by the motor must be equal to the sum of all frictional forces aiming to decelerate the car. And objects that are in free fall for a longer period of time reach and maintain a terminal velocity, implying that the gravitational force that pulls the body towards the surface (or rather center) of Earth is exactly balanced by air resistance.

Example 1:

A train of mass m = 80,000 kg is pushed by a motor force of F_1 = 95,000 N. When in motion, the train experiences a resistance force of F_2 = 28,000 N due to ground friction.

a) What is the resulting acceleration of the train? How fast does the train go after 10 seconds assuming it started at rest?

b) How do these values change when the train is filled with 130 people (each person weighing 75 kg)?

Solution:

a)

To calculate the acceleration, we need to determine the net force acting on the train. The net force is the difference of forces pushing and pulling on the train. Hence:

$F = F_1 - F_2$

$F = 95,000\ N - 28,000\ N = 67,000\ N$

Using Newton's second law we can now easily compute the acceleration of the train:

$F = m \cdot a$

$67,000\ N = 80,000\ kg \cdot a$

Divide both sides by 80,000:

$a = 67,000 / 80,000\ m/s^2 \approx 0.84\ m/s^2$

Assuming the train starts at rest, its speed is thus:

$0.84 \cdot 1\ m/s = 0.84\ m/s$ *after 1 second*

$0.84 \cdot 2\ m/s = 1.68\ m/s$ *after 2 seconds*

$0.84 \cdot 3\ m/s = 2.52\ m/s$ *after 3 seconds*

Continuing this train of thought (excuse the pun), we can conclude that the train's speed after 10 seconds is:

$v = 0.84 \cdot 10\ m/s = 8.4\ m/s \approx 19\ mph$

b)

To find the solution to b), we need to update the mass of the train. Since the 130 people weigh 130·75 kg = 9,750 kg, the mass of the train changes to m = 80,000 kg + 9,750 kg = 89,750 kg. The net force remains the same (67,000 N), so we can immediately turn to Newton's second law to find the acceleration.

$F = m \cdot a$

$67,000 \, N = 89,750 \, kg \cdot a$

Divide both sides by 89,750:

$a = 67,000 / 89,750 \, m/s^2 \approx 0.75 \, m/s^2$

Which is 11 % lower - a noticeable difference. Assuming the train starts at rest, its speed is thus:

$0.75 \cdot 1 \, m/s = 0.75 \, m/s$ *after 1 second*

$0.75 \cdot 2 \, m/s = 1.50 \, m/s$ *after 2 seconds*

$0.75 \cdot 3 \, m/s = 2.25 \, m/s$ *after 3 seconds*

And after 10 seconds:

$v = 0.75 \cdot 10 \, m/s = 7.5 \, m/s \approx 17 \, mph$

Example 2:

a) The Bugatti Veyron Super Sport, the fastest street-legal production car, can go from rest to 60 mph (27 m/s) in roughly 2.5 seconds. What is the acceleration experienced by the driver? What is the net force acting on the car during acceleration? The mass of the car is m = 1,900 kg.

b) We want to build a car that can go from rest to 60 mph (27 m/s) in just 2 seconds and has a mass of m = 2,000 kg. Compute the force our motor must be able to provide to achieve this given a resistance force of F_R = 8,500 N due to friction.

Solution:

a)

The acceleration a can be computed by dividing the change in speed by the elapsed time. This leads to:

$a = 27 / 2.5$ m/s² = 10.8 m/s² ≈ 1.1 g's

Which is a bit more than free fall - quite the thrill! To calculate the net force acting on the car during acceleration, we apply Newton's second law.

$F = m \cdot a$

$F = 1,900$ kg · 10.8 m/s² = 20,520 N

b)

Before we can compute the forces involved, we need to determine the acceleration. Since the change in speed is 27 m/s and the elapsed time 2 seconds, we get:

$a = 27 / 2 \ m/s^2 = 13.5 \ m/s^2$

With Newton's second law we can calculate the net force acting on the car during acceleration:

$F = m \cdot a$

$F = 2,000 \ kg \cdot 13.5 \ m/s^2 = 27,000 \ N$

The net force F is the difference of the motor force F_M and resistance force F_R: $F = F_M - F_R$. We can solve this for F_M and insert the corresponding values.

$F_M = F + F_R$

$F_M = 27,000 \ N + 8,500 \ N = 35,500 \ N$

As you can see, with the definition of acceleration and Newton's second law we can already do a lot of interesting physics with just a tiny bit of algebra. Now we are ready to have a look at the many different types of forces scientists have identified. We will start with the force that is most familiar to us, the gravitational force.

Gravitational Force

All objects exert a gravitational pull on all other objects. The Earth pulls you towards its center and you pull the Earth towards your center. Your car pulls you towards its center and you pull your car towards your center (of course in this case the forces involved are much smaller, but they are there). It is this force that invisibly tethers the Moon to Earth, the Earth to the Sun, the Sun to the Milky Way Galaxy and the Milky Way Galaxy to its local galaxy cluster.

Experiments have shown that the magnitude of the gravitational attraction between two bodies depends on their masses. If you double the mass of one of the bodies, the gravitational force doubles as well. The force also depends on the distance between the bodies. More distance means less gravitational pull. To be specific, the gravitational force obeys an inverse-square law. If you double the distance, the pull reduces to $1/2^2 = 1/4$ of its original value. If you triple the distance, it goes down to $1/3^2 = 1/9$ of its original value. And so on. These dependencies can be summarized in this neat formula:

$$F_G = G \cdot m \cdot M / r^2$$

With F being the gravitational force in Newtons, m and M the masses of the two bodies in kilograms, r the center-to-center distance between the bodies in meters and $G = 6.67 \cdot 10^{-11}$ N m^2 kg^{-2} the (cumbersome) gravitational constant. With this formula, that has first been derived at the end of the seventeenth century and has sparked an ugly plagiarism dispute between Newton and Hooke, you can calculate the gravitational pull between two objects for any situation.

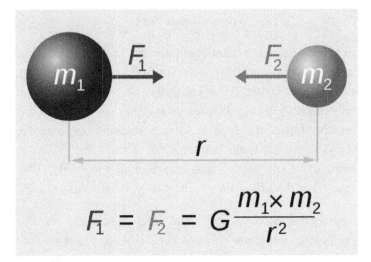

$$F_1 = F_2 = G\frac{m_1 \times m_2}{r^2}$$

(Gravitational attraction between two spherical masses)

If you have trouble applying the formula on your own or just want to play around with it a bit, check out the free web applet "Newton's Law of Gravity Calculator" that can be found on the website of the UNL astronomy education group. It allows you to set the required inputs (the masses and the center-to-center distance) using sliders that are marked special values such as Earth's mass or the distance Earth-Moon and calculates the gravitational force for you.

Example 3:

Calculate the gravitational force a person of mass m = 72 kg experiences at the surface of Earth. The mass of Earth is M = 5.97·10²⁴ kg and the distance from the center to the surface r = 6,370,000 m. Show that the acceleration the person experiences in free fall is roughly 10 m/s².

Solution:

To arrive at the answer, we simply insert all the given inputs into the formula for calculating gravitational force.

$F_G = G \cdot m \cdot M / r^2$

$F_G = 6.67 \cdot 10^{-11} \cdot 72 \cdot 5.97 \cdot 10^{24} / 6,370,000^2 \, N \approx 707 \, N$

So the magnitude of the gravitational force experienced by the m = 72 kg person is 707 N. In free fall, he or she is driven by this net force (assuming that we can neglect air resistance). Using Newton's second law we get the following value for the free fall acceleration:

$F = m \cdot a$

$707 \, N = 72 \, kg \cdot a$

Divide both sides by 72 kg:

$a = 707 / 72 \, m/s^2 \approx 9.82 \, m/s^2$

Which is roughly (and more exact than) the 10 m/s² we've been using in the introduction. Except for the overly small and large numbers involved, calculating gravitational pull is actually quite straight-forward.

As mentioned before, gravitation is not a one-way street. As the Earth pulls on the person, the person pulls on the Earth with the same force (707 N). However, Earth's mass is considerably larger and hence the acceleration it experiences much smaller. Using Newton's second law again and the value M = 5.97·10²⁴ kg for the mass of Earth we get:

$F = m \cdot a$

$707\ N = 5.97 \cdot 10^{24}\ kg \cdot a$

Divide both sides by 5.97·10²⁴ kg:

$a = 707 / 5.97 \cdot 10^{24}\ m/s^2 \approx 1.18 \cdot 10^{-22}\ m/s^2$

So indeed the acceleration the Earth experiences as a result of the gravitational attraction to the person is tiny.

Example 4:

By how much does the gravitational pull change when the person of mass m = 72 kg is in a plane (altitude 10 km = 10,000 m) instead of the surface of Earth? For the mass and radius of Earth, use the values from the previous example.

Solution:

In this case the center-to-center distance r between the bodies is a bit larger. To be specific, it is the sum of the radius of Earth 6,370,000 m and the height above the surface 10,000 m:

$r = 6,370,000\ m + 10,000\ m = 6,380,000\ m$

Again we insert everything:

$F_G = G \cdot m \cdot M / r^2$

$F_G = 6.67 \cdot 10^{-11} \cdot 72 \cdot 5.97 \cdot 10^{24} / 6,380,000^2\ N \approx 705\ N$

So the gravitational force does not change by much (only by 0.3 %) when in a plane. 10 km altitude are not much by gravity's standards, the height above the surface needs to be much larger for a noticeable difference to occur.

With the gravitational law we can easily show that the gravitational acceleration experienced by an object in free fall does not depend on its mass. All objects are subject to the same 10 m/s² acceleration near the surface of Earth. Suppose we denote the mass of an object by m and the mass of Earth by M. The center-to-center distance between the two is r, the radius of Earth. We can then insert all these values into our formula to find the value of the gravitational force:

$F_G = G \cdot m \cdot M / r^2$

Once calculated, we can turn to Newton's second law to find the acceleration a the object experiences in free fall. Using F = m·a and dividing both sides by m we find that:

$a = F_G / m = G \cdot M / r^2$

So the gravitational acceleration indeed depends only on the mass and radius of Earth, but not the object's mass. In free fall, a feather is subject to the same 10 m/s² acceleration as a stone. But wait, doesn't that contradict our experience? Doesn't a stone fall much faster than a feather? It sure does, but this is only due to the presence of air resistance. Initially, both are accelerated at the same rate. But while the stone hardly feels the effects of air resistance, the feather is almost immediately slowed down by the collisions with air molecules. If you dropped both in a vacuum tube, where no air resistance can build up, the stone and the feather would reach the ground at the same time! Check out an online video that shows this interesting vacuum tube experiment, it is quite enlightening to see a feather literally drop like a stone.

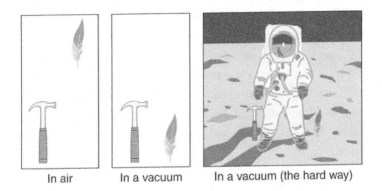

In air   In a vacuum   In a vacuum (the hard way)

(All bodies are subject to the same gravitational acceleration)

Since all objects experience the same acceleration near the surface of Earth and since this is where the everyday action takes place, it pays to have a simplified equation at hand for this special case. Denoting the gravitational acceleration by g (with $g \approx 10$ m/s^2) as is commonly done, we can calculate the gravitational force, also called weight, an object of mass m is subject to at the surface of Earth by:

$$F_G = m \cdot g$$

So it's as simple as multiplying the mass by ten. Depending on the application, you can also use the more accurate factor $g \approx 9.82$ m/s^2 (which I will not do in this book). Up to now we've only been dealing with gravitation near the surface of Earth, but of course the formula allows us to compute the gravitational force and acceleration near any other celestial body. I will spare you trouble of looking up the relevant data and do the tedious calculations for you. In the table below you can see what gravitational force and acceleration a person of mass m = 72 kg would experience at the surface of various celestial objects. The acceleration is listed in g's, with 1 g being equal to the free-fall acceleration experienced near the surface of Earth.

	Moon	Jupiter	Sun	Neutron Star
Grav. Force	113 N	1,765 N	19,700 N	70 trillion N
Acceleration	0.16 g's	2.5 g's	27.9 g's	100 billion g's

So while jumping on the Moon would feel like slow motion (the free-fall acceleration experienced is comparable to what you feel when stepping on the gas pedal in a common car), you could hardly stand upright on Jupiter as your muscles would have to support more than twice your weight. Imagine that! On the Sun it would be even worse. Assuming you find a way not get instantly killed by the hellish thermonuclear inferno, the enormous gravitational force would feel like having a car on top of you. And unlike temperature or pressure, shielding yourself against gravity is not possible.

What about the final entry? What is a neutron star and why does it have such a mind-blowing gravitational pull? A neutron star is the remnant of a massive star that has burned its fuel and exploded in a supernova, no doubt the most spectacular light-show in the universe. Such remnants are extremely dense - the mass of several suns compressed into an almost perfect sphere of just 20 km radius. With the mass being so large and the distance from the surface to the center so small, the gravitational force on the surface is gigantic and not survivable under any circumstances.

If you approached a neutron star, the gravitational pull would actually kill you long before reaching the surface in a process called spaghettification. This unusual term, made popular by the brilliant physicist Stephen Hawking, refers to the fact that in intense gravitational fields objects are vertically stretched and horizontally compressed. The explanation is rather straight-forward: since the strength of the gravitational force depends on the distance to the source of said force, one side of the approaching object, the side closer to the source, will experience a stronger pull than the opposite side. This leads to a net force stretching the object. If the gravitational force is large enough, this would make any object look like a thin spaghetti. For humans spaghettification would be lethal as the stretching would cause the body to break apart at the weakest spot (which presumably is just above the hips). So my pro-tip is to keep a polite distance from neutron stars.

Ground Friction

Another force we often encounter in our daily lives is ground friction. It is a result of uneven surfaces pushing against each other. Here's a quick experiment for you: put the tablet on which you are reading this book on the ground and slide it across with one hand. Now do this again, but this time press down on the tablet with your other hand while sliding it. What did you notice? When you press down on the tablet, you need to apply a greater force to slide it. So the magnitude of the frictional force F_R you must overcome to get the tablet moving obviously depends on the force at which the ground pushes on the object. The latter force is called the normal force F_N and is usually just a result of the gravitational pull on the object.

$F_R = \mu \cdot F_N$

The factor μ is a dimensionless quantity (meaning that it does not have any physical unit) called frictional coefficient and depends on the nature and state of the materials that are in contact. This quantity is very difficult to compute theoretically and thus has to be determined by experiments. When solving problems involving ground friction, you can look up a suitable value for μ in a table that summarizes experimental results.

For example, for wood-on-wood friction the coefficient takes on values between $\mu = 0.25 - 0.5$ (lubrication reduces this to roughly $\mu = 0.2$), while for concrete-on-wood friction the value is a bit higher, around $\mu = 0.62$. The frictional coefficient can also be greater than one, in which case the frictional force exceeds the normal force that the ground exerts on the object. One example is rubber-on-rubber friction with the corresponding value $\mu = 1.16$. For a thorough table of frictional coefficients, be sure to check out the website engineeringtoolbox.com.

Example 5:

Suppose you want to push a wooden box with mass $m = 60$ kg across a concrete floor. What force do you have to apply to keep the box moving? Use the value $\mu = 0.62$ for the frictional coefficient.

Solution:

First we have to determine the normal force F_N between the ground and the object, which in this case is equivalent in magnitude to gravitational pull the box experiences. We can use the simplified gravitational law for the special case of motion near the surface of Earth to calculate it:

$F_N = m \cdot g$

$F_N = 60 \ kg \cdot 10 \ m/s^2 = 600 \ N$

Now that we know the contact force, we can apply the ground friction formula to calculate the resistance force.

$F_R = \mu \cdot F_N$

$F_R = 0.62 \cdot 600 \ N = 372 \ N$

This is the force we have to apply to keep the box moving at constant speed. Just by the way: is 372 N a lot? Can you do this on your own or should you get your neighbor's help? Let's find out. What mass does this force correspond to in terms of weight? Setting up the equation:

$F_G = m \cdot g$

$372\ N = m \cdot 10\ m/s^2$

And dividing by 10 m/s² leads to:

$m = 372 / 10\ kg = 37.2\ kg$

So the frictional force we have to push against is equivalent to the weight of a 37.2 kg object - not impossibly large, but still a lot for one person. Seems like inviting your neighbor to a beer is a good idea after all (it always is).

Note that a similar formula is used for solving problems involving rolling rather than sliding. If you roll an object across the ground, the frictional force F_R you have to push against to keep the object rolling also grows proportionally to the normal force F_N. However, the frictional coefficient is replaced by the rolling resistance coefficient c, another a dimensionless quantity that depends on the materials in contact. Its value is generally much lower than the frictional coefficient.

$F_R = c \cdot F_N$

For railroad steel wheels on steel tracks the rolling resistance is between c = 0.001 - 0.002. The rolling resistance coefficient of a bicycle tire on dry asphalt is around double that, c = 0.004. For the special case of pneumatic car or truck tires on dry asphalt there is even a neat experimental formula to compute the rolling resistance coefficient from the tire's pressure p (in bar) and the speed v (in m/s) of the vehicle:

$$c = 0.005 + (1/p) \cdot (0.01 + 0.000{,}012 \cdot v^2)$$

There are several interesting things the formula can tell us. The coefficient of a tire is around 0.005 for speeds close to zero and increases considerably as the vehicle becomes faster. Assuming a pressure of p = 2.2 bar, the rolling resistance coefficient is c = 0.011 at a speed of v = 15 m/s (34 mph) and c = 0.014 at v = 30 m/s (68 mph), which is roughly triple the initial rolling resistance. One way to get less rolling resistance is to increase the pressure. At a speed of v = 15 m/s the coefficient is c = 0.011 at p = 2.2 bar and c = 0.010 at p = 2.4 bar, an almost 10 % reduction in rolling resistance. So it really pays to check the tire pressure.

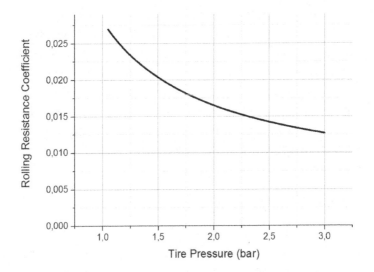

(Variation of rolling resistance coefficient with tire pressure at v = 33 m/s)

Example 6:

A person with mass $m_1 = 72$ kg is riding a bicycle with mass $m_2 = 12$ kg on an asphalt road. What's the rolling resistance the cyclist must overcome assuming a rolling resistance coefficient of $c = 0.004$? What acceleration can be achieved assuming the cyclists can provide the force $F_C = 40$ N?

Solution:

To calculate the rolling resistance, we need to determine the normal force that results from the gravitational pull on the bicycle plus person. With a total mass of $m = 72$ kg + 12 kg = 84 kg, this is approximately:

$$F_N = m \cdot g$$

$$F_N = 84 \text{ kg} \cdot 10 \text{ m/s}^2 = 840 \text{ N}$$

The rolling resistance is thus:

$$F_R = c \cdot F_N$$

$$F_R = 0.004 \cdot 840 \text{ N} = 3.4 \text{ N}$$

When applying a force of $F_C = 40$ N against the rolling resistance $F_R = 3.4$ N, the net force accelerating the bicycle is $F = F_C - F_R = 40$ N - 3.4 N = 36.6 N. The corresponding acceleration can be computed via Newton's second law:

$$F = m \cdot a$$

$$36.6 \text{ N} = 84 \text{ kg} \cdot a$$

Divide both sides by 84 kg:

$a = 36.6 / 84$ m/s² ≈ 0.44 m/s²

\-

Example 7:

a) What's the rolling resistance of a car with mass $m = 1,200$ kg when driving at highway speeds (75 mph = 33 m/s)?

b) How does that change if there are four people (each person weighing 72 kg) and luggage with mass 120 kg in the car?

Assume a tire pressure of $p = 2.2$ bar.

Solution:

a)

Let's compute the rolling resistance coefficient first. Using the formula given in the text we get:

$c = 0.005 + (1/p)·(0.01 + 0.000,012·v²)$

$c = 0.005 + (1/2.2)·(0.01 + 0.000,012·33²) ≈ 0.015$

Since c doesn't depend on the car's mass, we can use the same value for both cases. Onto the normal force. In the first case the mass is $m = 1,200$ kg and the corresponding gravitational pull:

$F_N = m·g$

$F_N = 1,200 \text{ kg} \cdot 10 \text{ m/s}^2 = 12,000 \text{ N}$

The rolling resistance is then:

$F_R = c \cdot F_N$

$F_R = 0.015 \cdot 12,000 \text{ N} = 180 \text{ N}$

Note: that's still less than half the frictional force of the m = 60 kg wooden box we pushed across the concrete floor.

b)

With four passengers and luggage, the mass of the car needs to be updated. The resulting mass is:

$m = 1,200 \text{ kg} + 4 \cdot 72 \text{ kg} + 120 \text{ kg} = 1,608 \text{ kg}$

This increases the normal force to:

$F_N = m \cdot g$

$F_N = 1,608 \text{ kg} \cdot 10 \text{ m/s}^2 = 16,080 \text{ N}$

And the rolling resistance to:

$F_R = c \cdot F_N$

$F_R = 0.015 \cdot 16,080 \text{ N} \approx 241 \text{ N}$

A very much noticeable 34 % increase in friction. This increase would a bit less pronounced if we changed the tire pressure to 2.4 bar to counter the additional mass. I leave this calculation up to you.

Motion on Inclined Planes

Up to know we've assumed that the ground the motion takes place on is flat. But a lot of the motion we encounter in everyday life takes place on inclined planes. In this case the gravitational force $F = m \cdot g$ splits up into two forces: one force F_D that accelerates the object down the inclined plane and another force that pushes the object on the surface of the plane (equivalent in magnitude to the normal force F_N). To calculate them, we need to know the angle θ the plane makes with the horizontal and put that into the sine or cosine function.

$F_D = m \cdot g \cdot \sin(\theta)$

$F_N = m \cdot g \cdot \cos(\theta)$

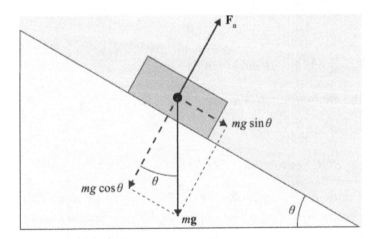

(Forces on an inclined plane)

Both Newton's second law and all the formulas for ground friction and rolling resistance remain valid. Note that both formulas given above reduce to common sense values for horizontal and vertical planes. Setting $\theta = 0°$ for the horizontal plane we get: $F_D = 0$ (no acceleration along the plane) and $F_N = m{\cdot}g$ (full gravitational push on the plane). And with $\theta = 90°$ for the vertical plane the formulas yield: $F_D = m{\cdot}g$ (free fall along the plane) and $F_N = 0$ (no push on the plane). So it all checks out.

Example 8:

We go back to the person with mass $m_1 = 72$ kg riding a bicycle with mass $m_2 = 12$ kg on an asphalt road. This time, the cyclists applies no force but rather let's the bicycle roll down an inclined plane that makes the angle $\theta = 4°$ with the horizontal.

a) What is the force accelerating the bicycle down the plane?

b) What is the rolling resistance of the bicycle? Assume a rolling resistance coefficient of $c = 0.004$.

c) What is the resulting acceleration?

Solution:

a)

The total mass of the bicycle plus person is $m = 72$ kg $+ 12$ kg $= 84$ kg. Inputting this into the formula used for determining the downwards force leads to:

$$F_D = m{\cdot}g{\cdot}sin(\theta)$$

33

$F_D = 84 \ kg \cdot 10 \ m/s^2 \cdot sin(4°)$

Using a calculator (set to degrees) we get:

$sin(4°) \approx 0.07$

Hence:

$F_D = 84 \ kg \cdot 10 \ m/s^2 \cdot 0.07 = 58.8 \ N$

b)

The normal force is:

$F_N = m \cdot g \cdot cos(\theta)$

$F_N = 84 \ kg \cdot 10 \ m/s^2 \cdot cos(4°)$

The calculator tells us that:

$cos(4°) \approx 1$

Thus:

$F_N - 84 \ kg \cdot 10 \ m/s^2 \cdot 1 \approx 840 \ N$

Now we can determine the rolling resistance:

$F_R = c \cdot F_N$

$F_R = 0.004 \cdot 840 \ N = 3.4 \ N$

As you might have noticed, we got the same result for the flat surface. This is because the angle is relatively small and the difference in rolling resistance thus not noticeable. For larger angles the difference becomes significant.

c)

To calculate the resulting acceleration we need to determine the net force acting on the bicycle. The net force is the difference of the force pushing the bicycle down the plane and the rolling resistance:

$F = F_D - F_R$

$F = 58.8 \, N - 3.4 \, N = 55.4 \, N$

As usual, we compute the acceleration using Newton's second law. This leads to:

$F = m \cdot a$

$55.4 \, N = 84 \, kg \cdot a$

Divide both sides by 84 kg:

$a = 55.4 \, / \, 84 \, m/s^2 \approx 0.66 \, m/s^2$

So computations on inclined planes require a bit more work, but the level of difficulty is not that much higher. It's mostly just a matter of knowing how to split up the gravitational force into one component along the plane (F_D) and another component perpendicular to the plane (F_N).

Before we move on, let's look at an interesting formula that can be derived using the formulas for the downward force F_D and normal force F_N presented above. As long as the angle of an inclined plane is very small, the normal force, and thus frictional force, is much greater than the downward force. In short: $F_R > F_D$ for small θ. Hence, unless pushed by an external force, the object will not move down the plane.

But as we increase the angle of the plane, making it steeper, the frictional force decreases while the downward force increases. At some critical angle θ' the two forces will take on the same value, in mathematical terms: $F_R = F_D$ at $\theta = \theta'$. Any further increase in steepness will cause the object to move down the plane without applying an external force. Let's derive a formula that allows us to compute the value of this critical angle. Using $F_R = \mu \cdot F_N$ and the formula for F_N for inclined planes we get:

$$F_R = \mu \cdot m \cdot g \cdot \cos(\theta)$$

We are interested in finding out at which angle this value is equal to the to the downwards push F_D. So let's set $F_R = F_D$ and see what we can do with the resulting equation:

$$F_R = F_D$$

$$\mu \cdot m \cdot g \cdot \cos(\theta') = m \cdot g \cdot \sin(\theta')$$

We can simplify this by dividing both sides by $m \cdot g$:

$$\mu \cdot \cos(\theta') = \sin(\theta')$$

Now we divide both sides by $\cos(\theta')$ and remember that the expression $\sin(\theta') / \cos(\theta')$ is equal to $\tan(\theta')$. This leads to:

$$\tan(\theta') = \mu$$

And voilà - a neat formula that allows us to quickly compute the critical angle θ', that is, the angle at which the downward push starts overtaking the frictional force and the object begins moving on its own. The formula even holds true for rolling motion, in which case we have to replace the frictional coefficient μ by the rolling resistance coefficient c. Note that the critical angle does not depend on the mass of the object or the gravitational acceleration. It doesn't matter if the object weighs one kilogram or one ton and it doesn't matter if the inclined plane is on Earth or Mars. The critical angle is solely determined by the nature and state of the materials in contact.

Example 9:

What's the critical angle at which a wooden box will begin sliding down an inclined plane made of concrete? Use μ = 0.62 for the frictional coefficient. Compare this to the angle at which a bicycle will start rolling down an asphalt plane without external forces pushing it. Set c = 0.004.

Solution:

Let's look at the wooden box on the concrete plane first. Using the formula derived above we get:

$tan(\theta') = \mu$

$tan(\theta') = 0.62$

Applying the inverse tangent (which can be found on almost all calculators - look for the button labeled tan⁻¹ or arctan and make sure that the calculator is set to degrees, not radians):

$\theta' = tan^{-1}(0.62) \approx 32°$

Once this angle is exceeded, the box will start sliding. Onto the bicycle on the asphalt plane. Using the same formula we get:

$tan(\theta') = c$

$tan(\theta') = 0.004$

Applying the inverse tangent:

$\theta' = tan^{-1}(0.004) \approx 0.2°$

So even a subtle slope will get the bicycle going.

Fluid Resistance

Another source of friction (between objects as well as between students and their teachers) is fluid resistance. This force is a result of an object colliding with the molecules of a fluid, usually air or water, in the object's path. An obvious factor is the speed v (in m/s) of the body relative to the fluid. The larger the speed, the higher the resistance force will be. Experiments and theoretical computations showed that rather than a simple linear proportionality, the resistance force grows with the square of the relative speed. This means that if you double the speed, the resistance force goes up by a factor of $2^2 = 4$, if you triple the speed, the resistance force increases by a factor $3^2 = 9$, and so on.

Other factors that influence fluid friction are the density D (in kg/m^3) of the fluid and the cross-sectional area A (in m^2) of the object relative to the direction of motion. Increase any of the two and the resistance force gets stronger. A dimensionless quantity that goes by the name drag coefficient c_D takes into account the geometrical shape of the body. The lower the value of the object's drag coefficient, the smaller the resistance force will be. Just like the frictional coefficient, the drag coefficient is very difficult to compute theoretically and thus has be determined via (wind tunnel) experiments. Hence, when solving problems involving fluid resistance, make sure to have a good table of drag coefficients at hand.

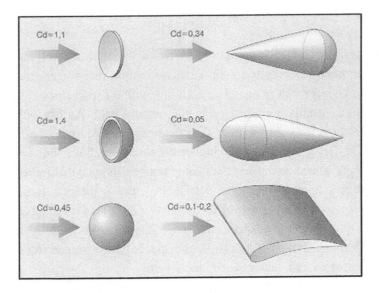

(Drag coefficients for common shapes)

What are some typical values for the drag coefficient? At the low end of the scale are the animals that live in the water and the man-made objects designed for high-speed travel. For example, a dolphin has a drag coefficient of $c_D = 0.004$, which is about as low as it gets. The drag coefficient of supersonic fighter jets is around $c_D = 0.015$, about four times higher than the dolphin, but still pretty low. The drag coefficient of cars can be as low as $c_D = 0.2$ for fly sports cars and as high as $c_D = 0.8$ for older cars such as the T-Ford. Common production cars are around $c_D = 0.3$. Unfortunately, people are not quite as streamlined as planes or cars. A person standing or a cyclist in upright position has a drag coefficient between $c_D = 1 - 1.3$. Among the few things that perform even worse in terms of fluid motion are plain rectangular boxes, which have a drag coefficient around $c_D = 2$. The box is about the most unfavorable shape you could give an object that is supposed to move through a fluid.

Let's turn to the formula that will allow us to compute the magnitude of the resistance force F_D (also referred to as drag). When all the quantities discussed above are known, calculating drag is a walk in the park. Just insert the numbers and evaluate:

$$F_D = 0.5 \cdot c_D \cdot D \cdot A \cdot v^2$$

Another formula often needed when solving problems involving fluid resistance is the barometric formula. With its help we can calculate, or rather approximate, the density of air D (in kg/m³) at altitude h (in m) above sea level. The letter "e" stands for Euler's constant e = 2.7182 ...

$$D = 1.25 \cdot e^{-0.000,1 \cdot h}$$

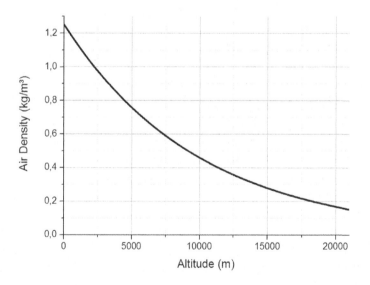

(Variation of air density with altitude)

Example 10:

A common car has a drag coefficient $c_D = 0.3$ and cross-sectional area $A = 2\ m^2$. Calculate the air resistance experienced by the car a) at sea-level and b) on a mountain road at altitude $h = 2,000\ m$. Assume highway speeds (75 mph = 33 m/s) for both cases.

Solution:

a)

First we need to determine the density of air at sea level. According to the barometric formula, its value is:

$$D = 1.25 \cdot e^{-0.000,1 \cdot h}$$

$$D = 1.25 \cdot e^{-0.000,1 \cdot 0} = 1.25 \cdot e^0 = 1.25\ kg/m^3$$

Now we can calculate the air resistance:

$$F_D = 0.5 \cdot c_D \cdot D \cdot A \cdot v^2$$

$$F_D = 0.5 \cdot 0.3 \cdot 1.25 \cdot 2 \cdot 33^2\ N \approx 408\ N$$

Note that this is much higher than the rolling resistance the car experiences at the same speed (roughly 180 N for a 1,200 kg car according to our calculations in example 7).

b)

Onto the mountain road. The barometric formula tells us that the air density at altitude $h = 2,000\ m$ is approximately:

$$D = 1.25 \cdot e^{-0.000,1 \cdot h}$$

$$D = 1.25 \cdot e^{-0.000,1 \cdot 2,000} = 1.25 \cdot e^{-0.2} \approx 1.02 \ kg/m^3$$

Resulting in the resistance force:

$$F_D = 0.5 \cdot c_D \cdot D \cdot A \cdot v^2$$

$$F_D = 0.5 \cdot 0.3 \cdot 1.02 \cdot 2 \cdot 33^2 \ N \approx 333 \ N$$

Still significantly higher than the rolling resistance, but an 18 % reduction compared to driving at sea level.

Example 11:

Resistance forces in water are much stronger than in air because of the higher density. Calculate the resistance force experienced by a submarine ($c_D = 0.02$ and $A = 12 \ m^2$) at a speed of $v = 5 \ m/s$. The density of water is $D = 1,000 \ kg/m^3$.

Solution:

Inputting the quantities leads to:

$$F_D = 0.5 \cdot c_D \cdot D \cdot A \cdot v^2$$

$$F_D = 0.5 \cdot 0.02 \cdot 1000 \cdot 12 \cdot 5^2 \ N = 3,000 \ N$$

This is eight times the resistance force encountered by the common car despite the fact that the submarine's drag coefficient and speed are significantly lower. At highway speeds the drag would grow to a mind-blowing 131,000 N, you'd need a powerful propulsion system to counter that.

The formula for fluid resistance has some great applications, one of which is calculating the terminal velocity in free fall. When you drop an object, initially the dominant factor is the gravitational force F_G accelerating the object at around 10 m/s². But as speed builds up, the air resistance F_D grows. After some time, the magnitude of the air resistance becomes equal to the gravitational pull, in short: $F_D = F_G$. The net force acting on the body is then zero, resulting in a constant velocity. The falling object will remain at this velocity until it hits the ground. To derive a handy formula for the terminal velocity, we simply set $F_D = F_G$ and solve for v.

$F_D = F_G$

$0.5 \cdot c_D \cdot D \cdot A \cdot v^2 = m \cdot g$

Solved for the velocity, this is:

$v = \text{sqrt} (2 \cdot m \cdot g / (c_D \cdot D \cdot A))$

Where "sqrt" stands for the square root. We can immediately draw a few general conclusions from the formula. For one, the terminal velocity grows with the square root of the mass. In other words: if you increase the mass by a factor $2^2 = 4$, the terminal velocity doubles, if you increase it by $3^2 = 9$, the terminal velocity triples, and so on. An increase in cross-sectional area has the opposite effect, which is why parachutes are so efficient in bringing down the speed of a falling object. While the formula does have a lot of inputs, the mathematics is not that difficult - insert and evaluate. Note that with a few adjustments, the formula can be used to find the maximum speed of a car. For some examples of this, check out the book "More Great Formulas Explained".

Example 12:

Calculate the terminal velocity of a skydiver of mass m = 72 kg falling in a head first position (c_D = 0.8 and A = 0.3 m²). Assume an air density of D = 1.2 kg/m³ for the calculation.

Solution:

Using the terminal velocity formula we get:

$$v = sqrt\ (2{\cdot}m{\cdot}g\ /\ (c_D{\cdot}D{\cdot}A)\ )$$

$$v = sqrt\ (2{\cdot}72{\cdot}10\ /\ (0.8{\cdot}1.2{\cdot}0.3)\ )\ m/s \approx 70\ m/s$$

Which is roughly 160 mph, more than twice the highway speed limit. Note that we assumed the air density to be close to its sea level value. At greater altitudes, the air is noticeably thinner and thus the terminal velocity larger. For example, at an altitude of h = 2,000 m the terminal velocity is around v ≈ 77 m/s ≈ 175 mph. And at an altitude of h = 4,000 m, this even grows to v ≈ 85 m/s ≈ 190 mph.

Buoyancy

When talking about fluids, there's another force worth mentioning. While some objects sink when placed in water, others are able to float in it, meaning that there must be a force balancing the gravitational pull. The calculation of this force, called buoyancy, dates back to Archimedes, a Greek scientists born in 287 BC. He recognized that the upward force is equivalent to the weight of the fluid that is displaced by the object and then, according to legend, screamed "Eureka" and ran naked through the streets of Syracuse.

(Just your regular scientist having his Eureka moment)

To put Archimedes' principle in the form of an easy-to-apply formula, we need to know how to convert from mass m (in kg) to volume V (in m³). Given the density D (in kg/m³) of the material, the mass can be computed from the volume using $m = D \cdot V$. For example, wood has a density around D = 700 kg/m³. Hence, a wooden rod of volume V = 0.02 m³, which corresponds to a height of 10 cm, width 20 cm and length 1 m, has a mass of:

$$m = D{\cdot}V = 700 \text{ kg/m}^3 \cdot 0.02 \text{ m}^3 = 14 \text{ kg}$$

Fair enough. Now suppose we submerge an object of volume V in a fluid of density D_F. The above formula tells us that the mass of the fluid displaced by the object is: $m_F = D_F{\cdot}V$. According to Archimedes' principle, the buoyancy F_B is equivalent to the weight of the displaced fluid. The law $F_G = m{\cdot}g$ allows us to compute the weight of an object from its mass. This leads to the following formula for the weight of the displaced fluid (and thus buoyancy):

$$F_B = m_F{\cdot}g = D_F{\cdot}V{\cdot}g$$

In words: buoyancy is the product of the fluid density, the volume of the submerged object and the gravitational acceleration $g \approx 10 \text{ m/s}^2$. At around 2250 years of age, this is one of the oldest laws of physics in existence. It comes from a time when people believed the Earth to be flat and lightning a sign from an angry god. So Archimedes' discovery was indeed a spectacular breakthrough.

When it comes to whether an object will sink or swim, looking at buoyancy alone is not sufficient. Rather, we are interested in the net force, the difference between the upward push of buoyancy and the downward pull of gravity. For this, let's rewrite the the gravitational law a bit. Suppose the density of the submerged body is D. Its mass is then m = D·V. Using this, the expression for its weight becomes:

$$F_G = m{\cdot}g = D{\cdot}V{\cdot}g$$

The net force F is the difference between the buoyancy and the object's weight: $F = F_B - F_G$. Comparing the two formulas, it is easy to see that it all comes down to the densities. If the density of the fluid is greater than the density of the object ($D_F > D$), then the buoyancy exceeds the gravitational pull ($F_B > F_G$) and the object swims. On the other hand, if the density of the fluid is smaller than the object's density ($D_F < D$), gravity wins the battle ($F_B < F_G$) and the body sinks. The volume of the object or the gravitational acceleration is not relevant when it comes to swimming or sinking.

Example 13:

We submerge a wooden rod of volume $V = 0.02$ m³ and density $D = 700$ kg/m³ in water ($D_F = 1000$ kg/m³).

a) Decide whether the rod floats or sinks.

b) Calculate the buoyancy and weight of the rod and calculate the rate at which it is accelerated upwards / downwards.

Solution:

a)

Since the density of water is greater than the density of wood, the rod will float on the surface.

b)

The buoyancy is:

$F_B = D_F \cdot V \cdot g$

$F_B = 1000 \cdot 0.02 \cdot 10 \; N = 200 \; N$

For the weight of the rod we get:

$F_G = D \cdot V \cdot g$

$F_G = 700 \cdot 0.02 \cdot 10 \; N = 140 \; N$

To calculate the upward acceleration we turn to our old friend Newton and his faithful companion, the second law. For this we need to know the net force acting on the rod and its mass. The net force is simply F = 200 N - 140 N = 60 N and the mass can be computed from the rod's density and volume:

$m = D \cdot V = 700 \; kg/m^3 \cdot 0.02 \; m^3 = 14 \; kg$

Now apply Newton's second law:

$F = m \cdot a$

$60 \; N = 14 \; kg \cdot a$

Divide both sides by 14 kg:

$a = 60 \, / \, 14 \; m/s^2 \approx 4.3 \; m/s^2$

That's around half the free fall acceleration (in the opposite direction though) - surprisingly large! Of course, this acceleration cannot be maintained for a longer period of time as the water's resistance force quickly builds up and the rod assumes a terminal velocity. The calculated value should thus be interpreted as the initial acceleration.

Before we move on, a note about density. Using Archimedes' principle, we were able to conclude that if the surrounding fluid is less dense than the object, the object will sink. So how is it possible that a ship with a hull made of steel (D = 8,000 kg/m³) can float on water? Shouldn't it sink like a stone? If the ship were entirely made of steel, it would indeed sink. But obviously the steel hull is just a very small part of the ship in terms of volume. Most of it is just plain air having a density of D = 1.25 kg/m³. What counts is the object's average density, which is just total mass divided by total volume. And with so much of the ship being air, its average density will naturally be much lower than the density of steel (or water).

Of course, this works both ways. Let's go back to our wooden rod. It floats because water has a higher density than wood. Now suppose that due to some reason, an underground explosion perhaps, a large number of air bubbles form in the water. The presence of the air bubbles lowers the average density of the water. If the number of bubbles grows and the water's average density goes below the density of the wood as a result of that, the rod will start to sink.

Lift

Here's another force that goes well with the topic fluids and one that you most definitely should be aware of: lift. It is what enables helicopters, planes and birds to get off and stay off the ground. To understand how lift arises, we first need to take a look at the quantity pressure. Pressure P (measured in Newtons per square meter = N/m²) is defined as the force applied per unit area. Accordingly, we can calculate it by dividing the force F (in N) acting on a surface by its total area A (in m²):

$$P = F / A$$

For example, consider a box with mass m = 20 kg sitting on a surface. According to the simplified gravitational law, the weight of the box is F_G = m·g = 20·10 N = 200 N. This is the force the box exerts on the ground. Given that the base area of the box is A = 0.25 m², this translates into a pressure of P = 200 N / 0.25 m² = 800 N/m². Of course we can also reverse the situation and use the law to calculate forces from pressure. If a pressure of P = 500 N/m² acts across a surface with area A = 0.1 m², the total force on the surface is F = P·A = 500·0.1 N = 50 N. So far so good.

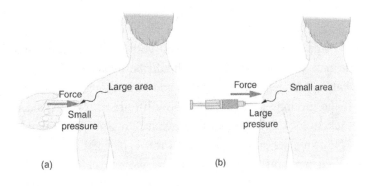

(a)

(b)

(Same force, different pressure)

Often forces arise as a difference of pressures across a surface. Suppose for example that on the right side of a surface the pressure is P_1, resulting in a force of $F_1 = P_1 \cdot A$ pushing on the surface from right to left. The pressure P_2 on the left side of the same surface gives rise to a force $F_2 = P_2 \cdot A$ pushing from left to right. The net force on the surface is the difference of the opposing forces: $F = F_2 - F_1 = (P_2 - P_1) \cdot A$. Usually the pressure difference $P_2 - P_1$ is symbolized by ΔP, which leads to:

$$F = \Delta P \cdot A$$

So force equals pressure difference times area. What does all of this have to do with lift? With the above formula, we're already half-way there. The only thing missing is Bernoulli's principle. It states the pressure of air decreases with velocity. So air moving slowly past a surface exerts a greater pressure on the surface than air flowing by at high speeds - just the opposite of what you would normally expect. With this in mind, we are ready to understand lift.

Consider a plane swooshing through the atmosphere. The wing of the plane splits up the air - some of it goes over the wing, the rest flows under it. The shape of the wing causes the air to move faster above the wing than below the wing. According to Bernoulli's principle, this means that the air moving below the wing exerts a greater pressure on the wing than the air moving above it. This pressure difference ΔP pushes the wing upwards and hence lift is born. Denoting the wing area by A, the lift can be calculated via $F_L = \Delta P \cdot A$.

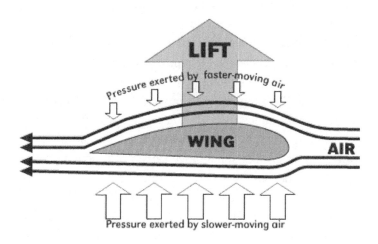

(Illustration of how lift is generated by an airfoil)

Since the pressure difference is rather difficult to determine theoretically, physicists usually turn to another, more practical, formula that is based on experimental results when calculating lift. Besides the wing area A (in m^2), the required inputs include the density D (in kg/m^3) of the surrounding fluid, the speed v (in m/s) of the wing relative to the fluid and the so-called lift coefficient c_L (dimensionless). Here the formula is in all its glory:

$F_L = 0.5 \cdot c_L \cdot D \cdot A \cdot v^2$

The formula looks quite similar to the one used to compute drag. Just like drag, lift increases linearly with the density of the fluid and quadratically with the relative speed, that is, double the relative speed, and the lift increases by a factor of four. However, in this case the area A refers to the wing area giving rise to the pressure difference and not the cross-sectional area that lies perpendicular to the flow.

The lift coefficient takes into account the geometric shape of the wing and its complex effect on the flow. You can find suitable values in tables summarizing experimental results for common wing shapes. For understanding certain aspects of flight it is worth knowing that the lift coefficient depends on the angle the wing's chord line makes with the incoming air. For small angles, the lift coefficient (and thus overall lift) increases as the angle between the wing and air flow grows. However, this simple mechanism of producing more lift only works up to a certain critical angle. For angles beyond the critical angle, the lift coefficient quickly decreases in value as the angle becomes larger. This sudden decrease in lift is called "stalling" and can be very dangerous for a plane and its passengers. Pilots are instructed to immediately lower the nose (and thus reduce the angle between the wing and air flow) when a stall occurs.

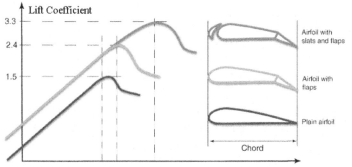

(Variation of lift coefficient with angle of attack)

Example 14:

The Cessna 172 has a wing area of A = 16.2 m² and lift coefficient around c_L = 1 (at moderate angles relative to the wind). What is the lift produced by the wings when taking off at sea level (D = 1.25 kg/m³) at a speed of v = 36 m/s? Compare this to the gravitational pull experienced by the Cessna given the take-off mass m = 1,000 kg.

Solution:

Since we have all the required inputs, we can simply plug them into the experimental formula and evaluate:

$$F_L = 0.5 \cdot c_L \cdot D \cdot A \cdot v^2$$

$$F_L = 0.5 \cdot 1 \cdot 1.25 \cdot 16.2 \cdot 36^2 \, N \approx 13,100 \, N$$

The opposing gravitational pull is:

$F_G = m \cdot g$

$F_G = 1,000 \cdot 10\ N = 10,000\ N$

Hence, the net force pushing the plane upwards is $F = F_L - F_G = 13,100\ N - 10,000\ N \approx 3,100\ N$ and the vertical acceleration $a = F / m \approx 3,100\ N / 1,000\ kg = 3.1\ m/s^2$.

Part of stable flight is balancing lift and weight. While the pilot can't do much to alter the weight of the plane, he or she is able to control lift by varying power (and thus speed) or changing the angle the wing makes with the wind. If the lift is smaller than the weight, the plane loses altitude, if the lift exceeds the weight, it gains altitude. For level flight, the lift must exactly balance the gravitational pull. With the formula above, we can easily compute the speed at which this occurs. Setting $F_L = F_G$ and solving for v leads to:

$v = \text{sqrt}\ (2 \cdot m \cdot g / (c_L \cdot D \cdot A)\ )$

This formula tells us that for level flight, the speed required grows with the square of the mass. Hence, if you increase the mass by a factor of $2^2 = 4$, the plane's speed must be doubled. The dependency of equilibrium speed on altitude is hidden in the factor fluid density. If the altitude of the plane increases, the density of the surrounding air decreases and this in turn means that a higher speed must be assumed to keep the lift and weight balanced. In short: the higher you go, the faster you must travel.

Example 15:

In level flight the Cessna's lift coefficient is roughly $c_L = 0.35$. Calculate the speed required to maintain an altitude of $h = 3,500$ m (use the barometric formula to estimate the corresponding air density). Use the values $m = 1,000$ kg for the mass and $A = 16.2$ m^2 for the wing area.

Solution:

Before we can use the level flight formula, we must compute the air density at altitude $h = 3,500$ m:

$$D = 1.25 \cdot e^{-0.000,1 \cdot h}$$

$$D = 1.25 \cdot e^{-0.000,1 \cdot 3,500} = 1.25 \cdot e^{-0.35} \approx 0.88 \ kg/m^3$$

With this done, we can insert the values:

$$v = sqrt \ (2 \cdot m \cdot g \ / \ (c_L \cdot D \cdot A) \ )$$

$$v = sqrt \ (2 \cdot 1,000 \cdot 10 \ / \ (0.35 \cdot 0.88 \cdot 16.2) \ ) \ m/s \approx 63 \ m/s$$

Which is around 140 mph. If you feel like it, you can confirm this result by calculating both the lift and weight and checking if they are indeed equal at this speed. You should get $F_L \approx F_G \approx 10,000$ N (the fact that we rounded the result introduces a slight error though).

Centrifugal Force

Enough with the fluids already! Let's go back to forces whose existence does not depend on the presence of a surrounding medium. One such force is the well-known centrifugal force. Surely you have felt it many times. It is the force that tries to push you and your car out of a curve. The magnitude of the centrifugal force F_C depends on three quantities: the mass m (in kg) of the body in motion, the speed v (in m/s) of said body and the radius r (in m) of the curve.

$F_C = m \cdot v^2 / r$

Note that again the dependence on velocity is quadratic: double the speed, and the strength of the force quadruples. However, a larger radius, and hence wider curve, decreases the magnitude of the outwards push. All vehicles must find a way to balance this force when going through a curve. In cars, this job is done by the sideways friction between the tires and the ground. Cyclists and bikers must lean into the curve to avoid getting carried away by the centrifugal force. And planes - we'll get to those in a moment.

Example 16:

Suppose you and your m = 1,000 kg car are going through a curve with radius r = 250 m at a speed of v = 28 m/s.

a) Calculate the magnitude of centrifugal force and compare this to the gravitational force on the car.

b) At what speed would your car be pushed out of the curve if your tires are able to produce a maximum sideways resistance force of 8,000 N?

Solution:

a)

This is simple - just plug in the values:

$F_C = m \cdot v^2 / r$

$F_C = 1,000 \cdot 28^2 / 250 \ N \approx 3,140 \ N$

The gravitational force acting on the car is:

$F_G = m \cdot g$

$F_G = 1,000 \cdot 10 \ N = 10,000 \ N$

So the magnitude of the centrifugal force is about one-third of the gravitational pull - you'll definitely feel that.

b)

Here must determine at what speed the centrifugal force reaches 8,000 N. We can set up the following equation to find the solution (I'll leave out the units for simplicity):

$F_C = 8,000$

$m \cdot v^2 / r = 8,000$

Insert the known quantities and simplify:

$1,000 \cdot v^2 / 250 = 8,000$

$4 \cdot v^2 = 8,000$

Divide both sides by 4:

$v^2 = 8,000 / 4 = 2,000$

Apply the square root and done:

$v = sqrt\ (2,000) \approx 45\ m/s \approx 100\ mph$

This is the speed at which the sideways push becomes too strong for the tires to counter it. If you are interested in more details on car performance in curves, with a focus on the dynamics of sliding and overturning, check out my e-book "More Great Formulas Explained".

Example 17:

Many science fiction movies feature rotating space stations in shape of a torus. The idea behind this is that the centrifugal force produced by the rotation would serve as artificial gravity. Suppose one such space station has a radius of r = 150 m. At what speed does the space station need to rotate for the centrifugal force on a person of mass m = 72 kg to be identical the gravitational pull on Earth?

(Illustration of a space station in form of a torus)

Solution:

We can take a similar approach as we did in part b) of the previous example. But before that, we need to compute the weight of the person at the surface of Earth:

$F_G = m \cdot g$

$F_G = 72 \cdot 10 \, N = 720 \, N$

We want the space station to produce the same centrifugal force on the person. So set up the corresponding equation:

$F_C = 720$

$m \cdot v^2 / r = 720$

Insert the known quantities and simplify:

$72 \cdot v^2 / 150 = 720$

$0.48 \cdot v^2 = 720$

Divide both sides by 0.48:

$v^2 = 720 / 0.48 = 1,500$

And apply the square root:

$v = sqrt (1,500) \approx 39 \ m/s \approx 87 \ mph$

Building a torus rotating at this speed is certainly feasible with our level of technology, so the idea of using the centrifugal outwards push for artificial gravity is not that far fetched. Note that if we use a different value for the person's mass, we would still arrive at the same rotation speed. Just like Earth's gravitational field, the centrifugal force causes all masses to accelerate at the same rate.

Roller coasters are quite the rush, especially if the track includes loops (though this authors prefers to stay away from them unless presented with a sufficient quantity of beer that will override the impulse). With the help of the centrifugal force, we can explain why a train doesn't simply fall off at the highest point in a loop and even calculate the required minimum speed of the train. For the train to complete the loop, the centrifugal force arising from the motion along the curved path needs to equal or exceed the train's weight at the top of the loop. Let's calculate the velocity at which the centrifugal force cancels the weight. Setting $F_C = F_G$ we get:

$m \cdot v^2 / r = m \cdot g$

Dividing both sides by m and multiplying by r brings us to the following neat formula for the critical speed:

$v = sqrt\ (r \cdot g)$

As long as the speed is equal to or greater than this critical value, the train will be able to go through the loop. Note that the mass of the train (and hence the number of people in the train) is irrelevant, it all comes down to the radius. The larger the radius of the loop, the more speed is required to complete it. Another relationship that is useful in this context is the formula linking the loop entry speed v_E with the speed at the top of the loop v:

$v_E = sqrt\ (v^2 + 4 \cdot r \cdot g)$

At this speed the train must enter the loop in order to achieve the critical speed when reaching the top. The formula can be derived with the help of the energy conservation principle (which for this application can be reduced to the formula: kinetic energy plus potential energy equals constant). However, we will not delve into this topic as the focus of the book shall remain on the quantity force.

Example 18:

Suppose we design a loop having a radius r = 15 m.

a) What is the critical speed at the top of the loop? What entry speed does this translate into?

b) Calculate the net force on a train of mass m = 1,500 kg at the top of the loop given the entry speed v_E = 30 m/s.

Solution:

a)

The critical speed at the top is:

$v = sqrt (r \cdot g)$

$v = sqrt (15 \cdot 10) \approx 12.2$ *m/s*

This leads to the corresponding entry speed:

$v_E = sqrt (v^2 + 4 \cdot r \cdot g)$

$v_E = sqrt (12.2^2 + 4 \cdot 15 \cdot 10) \approx 27.4$ *m/s*

b)

To calculate the forces acting on the train, we first need to determine the speed at the top of the loop. Using the given values and the formula linking the speed at the bottom and top of the loop, we can set up the following equation:

$v_E = sqrt (v^2 + 4 \cdot r \cdot g)$

$30 = sqrt (v^2 + 4 \cdot 15 \cdot 10)$

$30 = sqrt (v^2 + 600)$

Squaring both sides leads to:

$30^2 = v^2 + 600$

$900 = v^2 + 600$

Subtract 600 from both sides:

$300 = v^2$

Finally apply the square root:

$v = sqrt (300) \approx 17.3$ m/s

So when the entry speed is 30 m/s, the train will have the speed 17.3 m/s at the top (both safely beyond the required values). The net force at the top of the loop is the difference of centrifugal and gravitational force. For the centrifugal force we get the following value:

$F_C = m \cdot v^2 / r$

$F_C = 1,500 \cdot 17.3^2 / 15$ N $\approx 30,000$ N

The gravitational force acting on the train is:

$F_G = m \cdot g$

$F_G = 1,500 \cdot 10$ N $= 15,000$ N

This leads to the upwards net force:

$F = F_C - F_G = 30,000$ N - $15,000$ N $= 15,000$ N

Before we conclude the section on centrifugal force, let's take a quick look at what happens when a plane moves along a curved trajectory. Just like any other object in this situation, it will experience an outward push. However, unlike the car, there's no ground friction to counter the centrifugal force. So how does the plane balance the outward push? Well, they say a picture is worth a thousand words, so the image below should be quite enlightening.

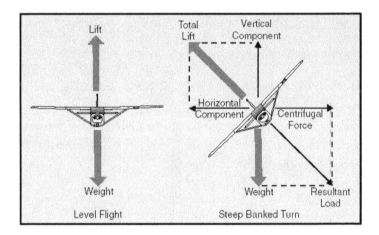

(Forces acting on a turning plane)

The lift vector always lies at a right angle to the wings that give rise to it. In level flight, the vector thus points in the opposite direction of the weight. However, as the plane banks, the lift vector banks along with it, meaning that it now has one vertical component and one horizontal component. For stable flight to occur, the vertical component of the lift must balance the weight and the horizontal component the centrifugal force.

Note that this requires the lift to be increased before (or at latest while) banking. If the lift balanced the weight before the turn was initiated, it won't do so during the turn as it must now balance both the unchanged weight and the newly created centrifugal force. For pilots this means: when turning, make sure to increase power. Otherwise the plane will most certainly lose altitude.

Spring Force

Springs are very simple and enormously useful devices. You can find them in clocks, suspension systems, electrical switches, cellular phones, measuring equipment, airplanes and even in spacecrafts. Hardly any machine, be it simple or complex, can be made to work without the use of springs. The most common type is the coil or helical spring. It consists of an elastic material, usually steel, formed into the shape of a helix. A key property of the spring is its ability to return to the original shape without damage after being deformed.

When deforming a spring, the spring pushes back with a force F_S that is roughly proportional to the displacement x (in m) from the equilibrium position. The force also depends on the so-called spring constant k (in N/m). This quantity can be interpreted as a measure of the spring's stiffness. To be more precise, it tells us how much force is necessary to deform the spring by one meter. The formula for the spring force is given by Hooke's law:

$$F_S = k \cdot x$$

More often than not, you'll find multiple springs used in connection rather than just one. In this case you have to compute the overall spring constant k of the system (also called equivalent spring constant) from the individual spring constants k_1, k_2, ... before applying the formula above. Usually the springs are either connected in series or in parallel. We will have a look at both arrangements.

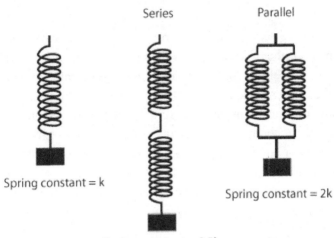

Series   Parallel

Spring constant = k

Spring constant = 0.5k

Spring constant = 2k

(Springs in series and parallel)

Suppose the springs are in series. Each spring in the system experiences the same force F and the total displacement x of the spring system is the sum of the individual displacements, so $x = x_1 + x_2 + ...$ Using Hooke's law we can rewrite this displacement equation in the following fashion:

$x = x_1 + x_2 + ...$

$x = F / k_1 + F / k_2 + ...$

$x = F \cdot (1 / k_1 + 1 / k_2 + ...)$

According to the same law, it also holds true that $x = F / k$. Equating the two expressions leads to the following general formula for the equivalent spring constant of springs in series:

$1 / k = 1 / k_1 + 1 / k_2 + ...$

Note that the overall spring constant in such systems is always smaller than the constant of any individual spring. In other words: the system as a whole is less stiff than the parts that make it up, a surprising result. If, for example, the arrangement consists of two springs of stiffness $k_1 = 1$ N/m and $k_2 = 2$ N/m, we get the equivalent spring constant:

$1 / k = 1 / 1 + 1 / 2$

$1 / k = 3 / 2$

$k = 2 / 3$ N/m ≈ 0.67 N/m

Indeed less stiff than any individual part. To simplify calculations it pays to keep one special case in mind. For a number of n identical springs of stiffness $k_1 = k_2 = ... = k_0$ in series, the formula produces the result $k = k_0 / n$, so the equivalent spring constant is equal to the spring constant of an individual spring divided by the number of springs in series.

Onto springs in parallel. In this case all the springs obviously experience the same displacement $x = x_1 = x_2 = ...$ According to Hooke's law, these displacements translate into the individual forces $F_1 = k_1 \cdot x$, $F_2 = k_2 \cdot x$, ... Since all the individual forces need to add up to the total force applied to the system, we get:

$F = F_1 + F_2 + ...$

$F = k_1 \cdot x + k_2 \cdot x + ...$

$F = (k_1 + k_2 + ...) \cdot x$

Comparing this with Hooke's law $F = k \cdot x$, we can immediately see that for springs in parallel, the overall spring constant is just the sum of the individual spring constants. Or to say that in mathematical terms:

$k = k_1 + k_2 + \ldots$

Here the situation is reversed as the equivalent spring constant is always greater than the constant of any individual spring, meaning that the entire system is stiffer than its parts. Our two springs of stiffness $k_1 = 1$ N/m and $k_2 = 2$ N/m would produce an overall stiffness $k = k_1 + k_2 = 1$ N/m + 2 N/m = 3 N/m when in parallel. If we use a number of n identical springs with stiffness $k_1 = k_2 = \ldots = k_0$ in parallel, the overall spring constant is just $k = k_0 \cdot n$, another special case to keep in mind.

Here's one more formula that will prove to be helpful for solving various problems involving springs. Suppose an object of mass m (in kg) hits a spring head-on with a speed of v (in m/s). The spring will decelerate this object while deforming. By how much will the spring be displaced from the equilibrium position before its restoring forces manage to bring the object to a complete halt? The following formula, which is a direct consequence of the conservation of energy, can help us answer that. Denoting the maximum displacement by X (in m), we get:

$X = v \cdot \text{sqrt} (m / k)$

Not surprisingly, the higher the speed and mass of the object, the larger the displacement will be. The relationship between maximum displacement and speed is linear, meaning that if you double the one, the other doubles as well. Armed with all of these formulas, we are able to solve a great variety of problems.

Example 19:

An object of mass m = 20 kg is attached to a spring. As a result of this, the spring is stretched by the distance x = 0.12 m.

a) Compute the spring constant.

b) How many springs of this type do you need to put in parallel to achieve an equivalent spring constant of k = 8500 N/m?

Solution:

a)

Before we can use Hooke's law, we have to convert the mass into a force. Using the law of gravitation, we get:

$F_G = m \cdot g$

$F_G = 20 \cdot 10 \, N = 200 \, N$

This is the force pulling on the spring. Now we can apply Hooke's law to determine the spring constant:

$F_S = k \cdot x$

$200 \, N = k \cdot 0.12 \, m$

Divide both sides by 0.12 m:

k = 200 N / 0.12 m ≈ 1,700 N/m

b)

When arranging springs in parallel, the individual spring constants add. With n identical springs of this type in the system, the equivalent spring constant k is thus:

k = 1,700 n/m + 1,700 N/m + ...

k = 1,700 N/m · n

Since we want k to be 8,500 N/m, we get:

8,500 N/m = 1,700 N/m · n

Divide by 1,700 N/m and done:

n = 8,500 / 1,700 = 5 springs

Example 20:

As a safety precaution, a roller coaster track is fitted with a spring-powered braking system that will decelerate the train in case the main brake fails. The overall spring constant of the system is k = 500 N/m, soft enough to keep the deceleration bearable if the train hits at full speed.

a) What is the stopping distance for a roller coaster train of mass m = 900 kg and speed v = 22 m/s?

b) The design requires the stopping distance to remain below x = 35 m. At what speed would the train reach this critical stopping distance?

b) The spring system consists of several helical steel springs in series, each having the spring constant k_0 = 3,500 N/m. How many of springs are needed for the braking system?

Solution:

a)

We can use the maximum displacement formula to determine the stopping distance X:

$X = v \cdot sqrt\ (m\ /\ k)$

$X = 22 \cdot sqrt\ (900\ /\ 500) \approx 29.5\ m$

b)

Here the maximum allowed displacement is given and the speed unknown. Let's insert the known quantities:

$X = v \cdot sqrt\ (m\ /\ k)$

$35 = v \cdot sqrt\ (900\ /\ 500)$

$35 \approx v \cdot 1.34$

Divide by 1.34:

$v \approx 35\ /\ 1.34 \approx 26.1\ m/s$

At speeds beyond the critical speed 26.1 m/s the stopping distance would exceed the design limit of 35 m.

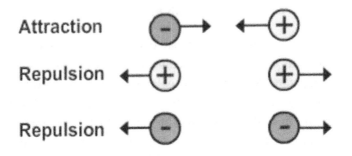

Attraction	(−)→ ←(+)
Repulsion	←(+) (+)→
Repulsion	←(−) (−)→

(Direction of Coulomb force for like and unlike charges)

Suppose we are given two particles carrying the charges q and Q. We'll denote the center-to-center distance between them by r (in m). To calculate the magnitude of the attractive or repulsive Coulomb force that acts on each particle, we can use the following formula (Coulomb's law):

$$F_Q = k \cdot q \cdot Q / r^2$$

Note that the Coulomb force diminishes with distance in the same inverse-square fashion as the gravitational force. If you double the distance between the charges, the magnitude of the force reduces to $1/2^2 = 1/4$ of its original value. If you triple the distance, it goes down to $1/3^2 = 1/9$ of its original value. Another similarity is the proportionality of the force to the product of the involved fundamental quantities. In simpler words: while the gravitational force is proportional to the product of the two masses, the Coulomb force is proportional to the product of the two charges. The factor k is called Coulomb's constant (no surprise there) and has the value $k = 8.99 \cdot 10^9 \ N \cdot m^2/C^2$.

Example 21:

Let's do a quick comparison to confirm that the gravitational force is indeed negligible on atomic scales. Calculate a) the Coulomb force and b) the gravitational force between two protons at a distance of $r = 2 \cdot 10^{-15}$ m (approximate size of the Helium nucleus). The mass of a proton is $m = 1.67 \cdot 10^{-27}$ kg and the charge is $q = 1.60 \cdot 10^{-19}$ C.

Solution:

All we need to do here is insert and evaluate, though the tiny numbers make this a bit more difficult than it sounds. I'll leave out the units for simplicity.

a)

For the Coulomb force we get:

$$F_Q = k \cdot q \cdot Q / r^2$$

$$F_Q = 8.99 \cdot 10^9 \cdot 1.60 \cdot 10^{-19} \cdot 1.60 \cdot 10^{-19} / (2 \cdot 10^{-15})^2$$

$$F_Q \approx 57.54 \, N$$

Since both protons are positively charged, the Coulomb force between them is repulsive, pushing the two protons away from each other. Note that because of the tiny mass of the proton, the acceleration resulting from this seemingly moderate force is gigantonormous. Using Newton's second law we get:

$$F = m \cdot a$$

$$57.54 \, N = 1.67 \cdot 10^{-27} \, kg \cdot a$$

Dividing both sides by the mass:

$a = 57.54 / 1.67 \cdot 10^{-27}$ m/s² ≈ $3.45 \cdot 10^{28}$ m/s²

This is a lot more than the gravitational acceleration one would experience at the surface of a neutron star.

b)

The gravitational pull is:

$F_G = G \cdot m \cdot M / r^2$

$F_G = 6.67 \cdot 10^{-11} \cdot 1.67 \cdot 10^{-27} \cdot 1.67 \cdot 10^{-27} / (2 \cdot 10^{-15})^2$

$F_G \approx 4.65 \cdot 10^{-35}$ N

This is much, MUCH weaker than the Coulomb force. So weak, that it's difficult to put into words. Let me try: if both forces were lengths and the gravitational force between the two protons were equivalent to one millimeter, the Coulomb force would be longer than the observable universe. That's how irrelevant gravitation is in the small-scale universe.

Example 22:

A particle of unknown charge q is placed at a fixed distance r = 4 m to a test charge Q = 0.002 C. A sensor measures the force between the charges to be $F_Q = 16,850$ N. Compute the charge q of the particle.

Solution:

We can set up an equation using Coulomb's law:

$F_Q = k \cdot q \cdot Q / r^2$

$16,850 = 8.99 \cdot 10^9 \cdot q \cdot 0.002 / 4^2$

$16,850 = 1,123,750 \cdot q$

Dividing by 1,123,750 leads to:

$q = 16,850 / 1,123,750 \approx 0.015\ C$

Our calculation in example 21 showed that there's are strong repulsive force between two protons. This is a problem. Why? As you might know, atomic cores are made up of neutrons and protons. While neutrons carry no electric charge and thus are not subject to the Coulomb force, protons are positively charged and repel each other. So how is it possible that they remain tightly packed in the nucleus? They should be pushing each other out of the atomic core at mind-blowing accelerations.

The reason why this doesn't happen is yet another force: the nuclear force (also referred to as the strong force). It is an enormously powerful attractive force that acts only on nuclear particles and only over very small distances. When you start pushing two protons together, initially you have to overcome their Coulomb repulsion. But when the distance between them becomes small enough, the nuclear force takes over and locks them into place. This is what makes the existence of atomic cores possible.

But back to the Coulomb force. It's worthwhile noting that unlike gravitation, the Coulomb force is influenced by the medium surrounding the charges. The value as calculated by the formula given above is only valid in vacuum, the presence of another medium reduces the strength of the force. Denoting by F_Q the magnitude of the Coulomb force for charges in a vacuum and F_{QM} its magnitude for charges in another medium, it holds true that:

$$F_{QM} = F_Q / \varepsilon_r$$

With $\varepsilon_r > 1$ being the relative permeability of the medium. The relative permeability of air is close to one ($\varepsilon_r = 1.0006$) and accordingly the difference is hardly noticeable. But water with its high relative permeability of $\varepsilon_r = 78$ significantly reduces the strength of the Coulomb force. So keep in mind that when it comes to electric attraction or repulsion, the surrounding medium must be taken into account.

Lorentz Force

Another force that plays an important role in the small-scale universe is the Lorentz force. It is the name given to the force that acts on a charged particle in a magnetic field. One special property of this force is that it always acts perpendicular to the direction of motion, giving the charge a sideways push and thus forcing it onto a circular path in the absence of other forces. This property also implies that the Lorentz force does not induce a change in velocity, it only changes the particle's direction. In addition to being perpendicular to the direction of motion, the force vector also lies at a right angle to the magnetic field lines.

For negative charges, the direction of the Lorentz force can be determined using the left hand rule. Point the index finger and thumb of your left hand as though you were imitating a pistol. Now put out your middle finger at a right angle to the index finger (see image below). If the negative charge moves in the direction of the index finger and the magnetic field is directed along the middle finger, the Lorentz force will act along the thumb. For positive charges the Lorentz force points in the opposite direction.

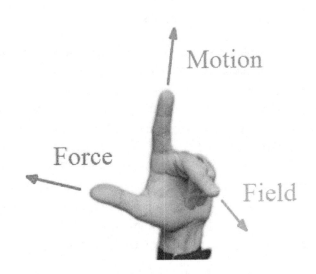

(Left-hand rule for negative charges)

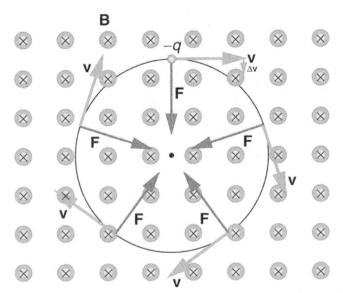

(Circular motion of a negatively charged particle in a magnetic field. The magnetic field lines point into the image)

Calculating the magnitude of the Lorentz force F_L is rather straight-forward. It is simply the product of the charge q (in C) carried by the particle, the strength B (in Tesla = T) of the magnetic field and the speed v (in m/s) of the particle. Double any of the quantities and the magnitude of the force experienced by the particle doubles as well.

$F_L = q \cdot B \cdot v$

Note that for particles at rest (v = 0 m/s), the magnitude of the Lorentz force is zero. Hence, the force only acts on moving charges, charges at rest are not bothered by the magnetic field. The strength of magnetic fields can vary over many orders of magnitude. The field produced by the electric activity within our brains is around 10^{-12} T, Earth's magnetic field is roughly fifty million times stronger than that ($5 \cdot 10^{-5}$ T). The strongest continuous magnetic field produced in a laboratory clocks in at 45 T, which is still far from the enormous magnetic fields found near neutron stars (up to 10^8 T).

Another useful formula for dealing with charges in magnetic fields can be derived by taking into account the centrifugal force. Since the charge remains on a circular path, the inward pull of the Lorentz force must be exactly balanced by the centrifugal outward push. In short: $F_L = F_C$. Inserting the corresponding formulas leads to a neat mathematical relationship between the speed v of the charged particle and the radius r (in m) of its trajectory in the magnetic field.

$F_L = F_C$

$q \cdot B \cdot v = m \cdot v^2 / r$

Solving for v, we get:

$$v = (q/m) \cdot B \cdot r$$

As usual, m (in kg) refers to the mass of the particle. The expression q/m is called the charge-to-mass ratio and has a value of $q/m = 1.76 \cdot 10^{11}$ C/kg for electrons and $q/m = 9.58 \cdot 10^7$ C/kg for protons. The above formula tells us that given a certain charge-to-mass ratio and field strength, a larger radius implies that the particle is moving at higher speeds.

Example 23:

Suppose a proton enters Earth's magnetic field ($B = 5 \cdot 10^{-5}$ T) and is forced onto a circular path with radius r = 2 m as a result of the electromagnetic interaction. Calculate the speed of the proton as well as the magnitude of the Lorentz force acting on it. The charge-to-mass ratio of a proton is $q/m = 9.58 \cdot 10^7$ C/kg and its charge $q = 1.60 \cdot 10^{-19}$ C.

Solution:

We can determine the speed of the proton using the formula derived from balancing the Lorentz and centrifugal force. Inserting the given values leads to:

$$v = (q/m) \cdot B \cdot r$$

$$v = 9.58 \cdot 10^7 \cdot 5 \cdot 10^{-5} \cdot 2 \ m/s \approx 9{,}600 \ m/s$$

With the speed known, calculating the Lorentz force is just a matter of applying the corresponding formula:

$$F_L = q \cdot B \cdot v$$

$$F_L = 1.60 \cdot 10^{-19} \cdot 5 \cdot 10^{-5} \cdot 9{,}600 \ N = 7.68 \cdot 10^{-20} \ N$$

A tiny force, but then again, the proton is a tiny object, so it will certainly feel the pull. And just like the Coulomb force, the Lorentz force is considerably stronger than gravitation on atomic scales.

The circular trajectory of a charged particle captured in a magnetic field is not stable. As noted in the introduction, acceleration does not just refer to changes in speed, but also to changes in direction of motion (vector nature of acceleration). Curved motion thus always implies acceleration, even if the speed of the object is remains the same. For our charged particle in the magnetic field this means that it is constantly undergoing acceleration due to the Lorentz force.

In the late nineteenth century it was discovered that charged particles lose energy whenever subject to acceleration. This phenomenon is known as Bremsstrahlung (German for "deceleration radiation"). The energy is given off in form of electromagnetic radiation, some of it visible to the naked eye, but most of it usually in the x-ray range. For non-relativistic particles, the intensity of the Lorentz force-induced radiation is proportional to the square of the magnetic field strength and particle velocity: $P \sim B^2 \cdot v^2$. Hence, stronger fields and faster particles translate into considerably higher levels of radiation. While this radiation makes it easier to track the particle's path in the magnetic field, it also means that the particle continuously loses energy, spiraling towards a center instead of remaining on its initial circular trajectory.

Casimir Force

The forces we have discussed so far are well-understood by the scientific community and are commonly featured in introductory as well as advanced physics books. In this section we will turn to a more exotic and mysterious interaction: the Casimir force. After a series of complex quantum mechanical calculations, the Dutch physicist Hendrick Casimir predicted its existence in 1948. However, detecting the interaction proved to be an enormous challenge as this required sensors capable picking up forces in the order of 10^{-15} N and smaller. It wasn't until 1996 that this technology became available and the existence of the Casimir force was experimentally confirmed.

So what does the Casimir force do? When you place an uncharged, conducting plate at a small distance to an identical plate, the Casimir force will pull them towards each other. The term "conductive" refers to the ability of a material to conduct electricity. For the force it plays no role though whether the plates are actually transporting electricity in a given moment or not, what counts is their ability to do so.

The existence of the force can only be explained via quantum theory. Classical physics considers the vacuum to be empty - no particles, no waves, no forces, just absolute nothingness. However, with the rise of quantum mechanics, scientists realized that this is just a crude approximation of reality. The vacuum is actually filled with an ocean of so-called virtual particles (don't let the name fool you, they are real). These particles are constantly produced in pairs and annihilate after a very short period of time. Each particle carries a certain amount of energy that depends on its wavelength: the lower the wavelength, the higher the energy of the particle. In theory, there's no upper limit for the energy such a particle can have when spontaneously coming into existence.

So how does this relate to the Casimir force? The two conducting plates define a boundary in space. They separate the space of finite extension between the plates from the (for all practical purposes) infinite space outside them. Only particles with wavelengths that are a multiple of the distance between the plates fit in the finite space, meaning that the particle density (and thus energy density) in the space between the plates is smaller than the energy density in the pure vacuum surrounding them. This imbalance in energy density gives rise to the Casimir force. In informal terms, the Casimir force is the push of the energy-rich vacuum on the energy-deficient space between the plates.

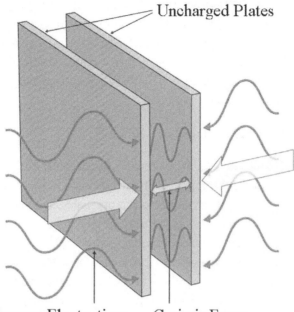

Uncharged Plates

Vacuum Fluctuations   Casimir Force

(Illustration of Casimir force)

It gets even more puzzling though. The nature of the Casimir force depends on the geometry of the plates. If you replace the flat plates by hemispherical shells, the Casimir force suddenly becomes repulsive, meaning that this specific geometry somehow manages to increase the energy density of the enclosed vacuum. Now the even more energy-rich finite space pushes on the surrounding infinite vacuum. Trippy, huh? So which shapes lead to attraction and which lead to repulsion? Unfortunately, there is no intuitive way to decide. Only abstract mathematical calculations and sophisticated experiments can provide an answer.

We can use the following formula to calculate the magnitude of the attractive Casimir force F_{CS} between two flat plates. Its value depends solely on the distance d (in m) between the plates and the area A (in m²) of one plate. The letters h = $6.63 \cdot 10^{-34}$ m² kg/s and c = $3.00 \cdot 10^8$ m/s represent Plank's constant and the speed of light.

$$F_{CS} = \pi \cdot h \cdot c \cdot A / (480 \cdot d^4) \approx 1.3 \cdot 10^{-27} \cdot A/d^4$$

Note that because of the exponent, the strength of the force goes down very rapidly with increasing distance. If you double the size of the gap between the plates, the magnitude of the force reduces to $1/2^4 = 1/16$ of its original value. And if you triple the distance, it goes down to $1/3^4 = 1/81$ of its original value. This strong dependence on distance and the presence of Plank's constant as a factor cause the Casimir force to be extremely weak in most real-world situations.

Example 24:

a) Calculate the magnitude of the Casimir force experienced by two conducting plates having an area A = 1 m² each and distance d = 0.001 m (one millimeter). Compare this to their mutual gravitational attraction given the mass m = 5 kg of one plate.

b) How close do the plates need to be for the Casimir force to be in the order of unity? Set $F_{CS} = 1$ N.

Solution:

a)

Inserting the given values into the formula for the Casimir force leads to (units not included):

$F_{CS} = 1.3 \cdot 10^{-27} \cdot A/d^4$

$F_{CS} = 1.3 \cdot 10^{-27} \cdot 1 / 0.001^4$

$F_{CS} \approx 1.3 \cdot 10^{-15} \, N$

Their gravitational attraction is:

$F_G = G \cdot m \cdot M / r^2$

$F_G = 6.67 \cdot 10^{-11} \cdot 5 \cdot 5 / 0.001^2$

$F_G \approx 0.0017 \, N$

This is more than a trillion times the magnitude of the Casimir force - no wonder this exotic force went undetected for so long. I should mention though that the gravitational force calculated above should only be regarded as a rough approximation as Newton's law of gravitation is tailored to two attracting spheres, not two attracting plates.

b)

Setting up an equation we get:

$F_{CS} = 1.3 \cdot 10^{-27} \cdot A/d^4$

$1 = 1.3 \cdot 10^{-27} \cdot 1 / d^4$

Multiply by d^4:

$d^4 = 1.3 \cdot 10^{-27}$

And apply the fourth root:

$d \approx 2 \cdot 10^{-7}$ m = 200 nanometers

This is roughly the size of a common virus and just a bit longer than the wavelength of violet light.

The existence of the Casimir force provides an impressive proof that the abstract mathematics of quantum mechanics is able to accurately describe the workings of the small-scale universe. However, many open questions remain. Quantum theory predicts the energy density of the vacuum to be infinitely large. According to Einstein's theory of gravitation, such a concentration of energy would produce an infinite space-time curvature and if this were the case, we wouldn't exist. So either we don't exist (which I'm pretty sure is not the case) or the most powerful theories in physics are at odds when it comes to the vacuum.

Part Two

Force - A Broader Understanding

I've been lying to you. Remember how I told you that Newton's second law states that force equals mass times acceleration? Well, it's a lie. Or at least not the whole truth. What we've been using so far is a simplified version of Newton's second law, a version, that only holds true if the mass of the object in motion is constant. If the mass is variable, $F = m \cdot a$ becomes invalid. The simplified version also prevented us from having a look at one of the most interesting forces out there: thrust. We'll get to that in a moment, but first, let's look at how Newton actually defined the quantity force.

For this we need to turn to another physical quantity called momentum. The definition of momentum is rather painless. An object with mass m (in kg) and velocity v (in m/s) is assigned the momentum: $p = m \cdot v$ (in kg m/s). So it's as simple as multiplying the mass by the velocity. For example, a car with mass m = 1,000 kg moving with v = 30 m/s has a momentum of $p = 1,000$ kg $\cdot$ 30 m/s = 30,000 kg m/s. Fair enough.

For reasons unknown, the total amount of momentum in the universe always remains constant. In other words: if one object loses momentum, for example because of mass loss or a decrease in speed, other objects have to gain exactly this amount of momentum in the opposite direction. Not a single process has been observed, on Earth or in deep space, that does not follow this rule. Conservation of momentum is how motion in our universe works.

Hence, it is not surprising that Newton turned to such a fundamental quantity when defining force. Without further ado, here's the original and most general form of Newton's second law. It states that force is the rate of change in momentum with respect to time. In other words: if the momentum changes by Δp over the time span Δt, the corresponding force is:

$$F = \Delta p / \Delta t$$

Now suppose the mass of the object that is in motion is constant. This means that a change in momentum Δp can only occur as a result of a the change in velocity Δv. According to the definition of momentum, it holds true that $\Delta p = m \cdot \Delta v$. If we divide this change in momentum by the elapsed time Δt (thus computing the force involved), we get:

$$F = \Delta p / \Delta t = m \cdot \Delta v / \Delta t$$

On the very right we now have the change in velocity divided by the change in time. Isn't that the definition of acceleration? It sure is. Go back to the very first formula in this book if you would like to confirm it. Hence, if the mass of the object doesn't change over time, Newton's general definition of force as the rate of change in momentum reduces to the formula $F = m \cdot a$. So what you learned was not wrong after all, but it was only a special case of a broader definition. Once the mass becomes a variable (which luckily doesn't happen too often), we have to go back to the general definition.

Since the mathematics behind Newton's second law in its

general form can get rather tricky and requires concepts from the field of calculus, we will not pursue this any further in this book. But you should keep the principle of momentum conservation in mind, it is not just vital for understanding how thrust is created, but also provides a deep insight into the workings of our universe. The same is true for other conservation laws such as mass-energy conservation (can't separate the two) or charge conservation. These "chosen" quantities are conserved during any process in the universe without there being an obvious why. It's just what the universe does.

Thrust

Suppose you are sitting in a boat that is at rest relative to the surrounding water and you want to get it moving. There's no paddle, but luckily you have a bag full of stones with you. What to do? Well, throw a stone and the boat will move in the opposite direction. How so? When throwing the stone (having the mass m and final speed v), you equip it with the momentum $p = m \cdot v$ in a certain direction. To conserve the momentum in the universe, someone or something has to gain the same amount of momentum in the opposite direction. This someone or something will obviously be you and the boat (with the combined mass m' and speed v' after the stone is thrown). Thus, it must hold true that $p' = p$ or $m' \cdot v' = m \cdot v$. For example, if the mass of the stone is $m = 5$ kg and it is thrown at $v = 10$ m/s from the boat with mass m' = 80 kg, we get:

$$m' \cdot v' = m \cdot v$$

$$80 \text{ kg} \cdot v' = 5 \text{ kg} \cdot 10 \text{ m/s}$$

$$v' \approx 0.63 \text{ m/s}$$

So given these values, the stone will cause the boat to move with 0.63 m/s. This is all well and good, but what does it have to do with thrust? Simple: rockets accelerate using the same principle, with the exception that they prefer to throw gas particles rather than stones out of the nozzle (though in theory, stones would work as well). Each gas particle carries a certain amount of momentum that the rocket must gain in the opposite direction for the momentum to be conserved.

Suppose your rocket expels Δm kilograms of fuel at a constant exhaust speed v. The total amount of momentum the rocket gains is $\Delta p = \Delta m \cdot v$. If this happens in the time span Δt, the force acting on the rocket must be $F = \Delta p / \Delta t = \Delta m \cdot v / \Delta t$ according to Newton's second law. Note that this expression contains the ratio $\Delta m / \Delta t$, which is just the rate at which the fuel is expelled through the nozzle. Denoting this mass flow rate by r (in kg/s), the formula for the resulting force takes on a very simple form:

$$F_T = r \cdot v$$

Long story short: the force the rocket experiences (also called thrust) is simply the product of the mass flow rate and the exhaust velocity, an elegant and powerful result we couldn't have gotten without resorting to the conservation of momentum and the generalized form of Newton's second law. There's one more thing to keep in mind though. The gas is leaving the nozzle at a pressure P_g that is significantly above the ambient pressure P_a outside the rocket. In other words: the pressure behind the rocket is greater than the pressure in front of it. As you know from the section on lift, pressure differences always give rise to a force directed from the high-pressure to the low-pressure region. So there must be an additional component to thrust, namely the force $F = \Delta P \cdot A$ pushing the rocket as a result of this pressure difference $\Delta P = P_g - P_a$.

$$F_T = r \cdot v + (P_g - P_a) \cdot A$$

Here the quantity A denotes the exit area of the nozzle. I should mention that physicists often prefer an alternative way of expressing the above formula. Instead of using the actual exhaust velocity v in their calculations, they work with the so-called effective exhaust velocity v_e defined by the equation:

$$v_e = v + (P_g - P_a) \cdot A / r$$

The general thrust formula then becomes:

$$F_T = r \cdot v_e$$

As you can confirm by inserting the expression for v_e into the formula. Another formula that is enormously useful when discussing rocket motion is Tsiolkovsky's rocket equation. It can be derived from the thrust equation using the awesome power of calculus and allows us to calculate the final speed v_f of a rocket from its initial mass m_0, initial speed v_0 and final mass m_f. Note that the rocket equation is only strictly valid in deep space (no atmosphere, no gravitation) and for non-relativistic speeds (speeds small compared to speed of light). The operator "ln" refers to the natural logarithm.

$$v_f = v_0 + v_e \cdot \ln(m_0 / m_f)$$

You can find the relativistic version of Tsiolkovsky's rocket equation, along with a more in-depth introduction to thrust and an overview of common and less common propulsion systems, in my e-book "Antimatter Propulsion" (suitable for beginners). Now let's turn to the examples.

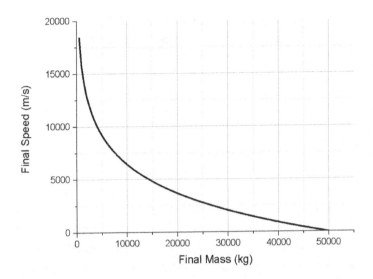

(Variation of final speed with final mass for a rocket with v_e = 4,000 m/s starting at rest v_0 = 0 m/s with mass m_0 = 50,000 kg)

Example 25:

Liquid-propellant rockets typically achieve a real exhaust velocity of around v = 3,600 m/s. Suppose our rocket expels gas at a rate of r = 350 kg/s and a pressure P_g = 275,000 N/m² through a nozzle having the exit area A = 0.6 m².

a) What is the effective exhaust velocity and thrust the rocket achieves in space (ambient pressure P_a = 0 N/m²)?

b) What are the corresponding values when the rocket operates at sea level (ambient pressure $P_a \approx 100,000$ N/m²)?

100

c) The rocket has a mass of m = 45,000 kg. What is the net force and acceleration the rocket experiences at launch from sea level? The thrust acts vertically upwards.

Solution:

a)

The effective velocity in space is:

$v_e = v + (P_g - P_a) \cdot A / r$

$v_e = 3,600 + (275,000 - 0) \cdot 0.6 / 350 \approx 4,070$ m/s

The resulting thrust is:

$F_T = r \cdot v_e$

$F_T = 350 \cdot 4,070 \approx 1,425,000$ N

b)

Let's see how the presence of a non-zero ambient pressure influences these values. The effective velocity at sea level is:

$v_e = v + (P_g - P_a) \cdot A / r$

$v_e = 3,600 + (275,000 - 100,000) \cdot 0.6 / 350 = 3,900$ m/s

This leads to the thrust:

$F_T = r \cdot v_e$

$F_T = 350 \cdot 3,900 = 1,365,000$ N

This is roughly 4 % lower than the achievable thrust in space - not a game-changer, but certainly something that

has to be taken into account by the engineers.

c)

The gravitational pull on the rocket is:

$$F_G = m \cdot g$$

$$F_G = 45,000 \cdot 10 \, N = 450,000 \, N$$

The net force is the difference of thrust and weight:

$$F = F_T - F_G$$

$$F = 1,365,000 - 450,000 \, N = 915,000 \, N$$

According to Newton's second law (the simplified form), the initial acceleration of the rocket is:

$$F = m \cdot a$$

$$915,000 \, N = 45,000 \, kg \cdot a$$

Divide by 45,000 kg:

$$a \approx 20 \, m/s^2 \approx 2 \, g's$$

So the astronauts in the rocket feel as though they are pushed into their seats at twice their weight. By the way: why did I suddenly use the simplified form of Newton's law again? Doesn't the mass of the rocket change over time (it's losing 350 kg each second), which makes the simplified form invalid? Indeed. But note the word "initial". The computed acceleration value only holds true at the moment of launch. To calculate the acceleration at a later time, we would be forced to use the general form of Newton's second law.

Example 26:

We want to launch a rocket with total mass m = 80,000 kg from sea level (ambient pressure $P_a \approx 100,000$ N/m²). Our propulsion system is able to expel the exhaust gas at a velocity of v = 3,800 m/s and a pressure P_g = 320,000 N/m². The exit area of the nozzle is A = 0.5 m². What mass flow rate is required in order for the thrust to balance the weight of the rocket?

Solution:

The weight of the rocket is:

$F_G = m \cdot g$

$F_G = 80,000 \cdot 10$ N $= 800,000$ N

We want to choose the mass flow rate r so that the thrust takes on the same value F_T = 800,000 N. For this we can do without the concept of the effective exhaust velocity and use the equation:

$F_T = r \cdot v + (P_g - P_a) \cdot A$

Inserting the known quantities leads to:

$800,000 = r \cdot 3,800 + (320,000 - 100,000) \cdot 0.5$

$800,000 = r \cdot 3,800 + 110,000$

Subtracting 110,000 we get:

$690,000 = r \cdot 3,800$

Finally divide by 3,800 and done:

$r = 690,000 / 3,800 \approx 182$ *kg/s*

Example 27:

A rocket having the effective exhaust velocity v_e = 4,500 m/s begins accelerating at an initial mass of m_0 = 65,000 kg and initial speed v_0 = 1,200 m/s. After burning all the available fuel, the mass of the rocket has decreased to m_f = 12,000 kg.

a) Calculate the final speed of the rocket.

b) Suppose the rocket burned its fuel at a rate of 65 kg/s. How long did the maneuver take? What was the average acceleration during this time?

Solution:

a)

We'll turn to Tsiolkovsky's rocket equation to solve the problem. Since all the required inputs are given, this should be a walk in the park.

$v_f = v_0 + v_e \cdot ln(m_0 / m_f)$

$v_f = 1,200 + 4,500 \cdot ln(65,000 / 12,000)$

$v_f \approx 8,800$ *m/s*

b)

This one we can solve with pure logic. The total amount of fuel burned during the maneuver is 65,000 kg - 12,000 kg = 53,000 kg. At a rate of 65 kg/s, this takes the time:

$t = 53,000 / 65\ s = 815\ s$

The average acceleration a can be computed by dividing the change in speed by the elapsed time. The change in speed is 8,800 m/s - 1,200 m/s = 7,600 m/s, thus:

$a = \Delta v / \Delta t$

$a = 7,600 / 815\ m/s^2 \approx 9.3\ m/s^2$

Beyond One Dimension

Up to now our discussion has mostly been one-dimensional, with forces and counter-forces acting along one line: forwards / backwards, left / right, up / down, inwards / outwards. In all these cases, computing the net force was just a matter of adding or subtracting two numbers. Unfortunately, many forces in real life are not aligned so neatly. When pushing a box for example, the force you apply is probably not completely horizontal, it will have both a horizontal and vertical component. To take this into account, we need to move beyond one dimension and the most elegant way of doing this is to make use of the vector notation.

Coordinate systems consist of three mutually perpendicular axis: the x-axis, y-axis and z-axis. We usually think of the x- and y-axis as spawning the ground (the horizontal) and the z-axis defining the vertical. When a real-life force acts on a body, it can have an x-component, y-component and z-component. We'll denote the corresponding component forces by F_x, F_y and F_z and summarize them in a vector:

$\mathbf{F} = (F_x, F_y, F_z)$

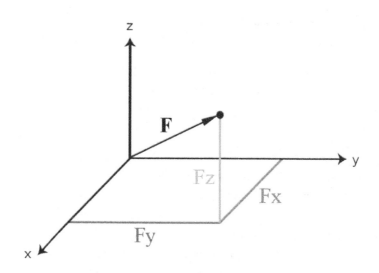

(A force vector in a three dimensional coordinate system)

For example, the vector **F** = (20 N, 0 N, -100 N) tells us that the body experiences a 20 N force in x-direction, no force in y-direction and a 100 N force opposite the z-direction (in other words: downwards). Net forces can be computed by adding vectors, which is quite simple. Suppose these two forces act on a certain body:

F₁ = (20 N, 0 N, -100 N)

F₂ = (-50 N, 10 N, 100 N)

To calculate the net force resulting from these forces, we just add the components separately.

F = **F**₁ + **F**₂

$\mathbf{F} = (20 \text{ N}, 0 \text{ N}, -100 \text{ N})$

$+ (-50 \text{ N}, 10 \text{ N}, 100 \text{ N})$

$\mathbf{F} = (20 - 50 \text{ N}, 0 + 10 \text{ N}, -100 + 100 \text{ N})$

$\mathbf{F} = (-30 \text{ N}, 10 \text{ N}, 0 \text{ N})$

So overall, the body only experiences a horizontal push: 30 N opposite the x-direction and 10 N in y-direction. The vertical components canceled each other out. This approach also works for three or more forces, we simply keep adding them component by component to find the net force. Can we also compute the corresponding acceleration? We sure can. To do this, we divide each component of the net force vector by the object's mass to get the acceleration vector.

$\mathbf{a} = \mathbf{F} / m = (F_x / m, F_y / m, F_z / m)$

By the way: mass (just like temperature, energy and charge) cannot have direction. Such quantities are the opposite of vectors and go by the name of scalars. But back to the acceleration. If the body experiencing the above net force $\mathbf{F} = (-30 \text{ N}, 10 \text{ N}, 0 \text{ N})$ has a mass of $m = 5$ kg, the resulting acceleration is:

$\mathbf{a} = \mathbf{F} / m$

$\mathbf{a} = (-30 \text{ N} / 5 \text{ kg}, 10 \text{ N} / 5 \text{ kg}, 0 \text{ N} / 10 \text{ kg})$

$\mathbf{a} = (-6 \text{ m/s}^2, 2 \text{ m/s}^2, 0 \text{ m/s}^2)$

So it is accelerated with 6 m/s² opposite the x-direction, 2 m/s² in y-direction and has no acceleration in the vertical direction. This is how Newton's second law works in three dimensions. Given this acceleration vector, what is the overall acceleration (the magnitude) experienced by the body? Is it simply -6 m/s² + 2 m/s² + 0 m/s² = -4 m/s², the sum of the components? Unfortunately, it's a bit more complicated than that. To find the magnitude of a vector, we have to add the squares of the components and then apply the square root. So the vector $\mathbf{F} = (F_x, F_y, F_z)$ has the magnitude:

$F = |\mathbf{F}| = $ sqrt $(F_x^2 + F_y^2 + F_z^2)$

Let's calculate the magnitude of the net force vector and the acceleration vector we used above:

$\mathbf{F} = $ (-30 N, 10 N, 0 N)

$F = $ sqrt $((-30)^2 + 10^2 + 0^2)$ N ≈ 31.6 N

$\mathbf{a} = $ (-6 m/s², 2 m/s², 0 m/s²)

$a = $ sqrt $((-6)^2 + 2^2 + 0^2)$ N ≈ 6.3 m/s²

You should definitely keep this technique in mind. Another formula that is helpful for interpreting vectors is the formula used for finding the angle the vector makes with the horizontal. This angle has a positive value if the vector points upwards and a negative value if it points downwards. Given the vector $\mathbf{F} = (F_x, F_y, F_z)$, the corresponding angle θ is:

$\tan(\theta) = F_z / $ sqrt $(F_x^2 + F_y^2)$

For example, suppose the velocity of an object is given by the vector $\mathbf{v} = $ (2 m/s, -2 m/s, 6 m/s), so it moves at a speed of 2 m/s in x-direction, 2 m/s opposite the y-direction and 6 m/s in vertical direction. Try picturing this motion in your mind. The angle the velocity vector makes with the horizontal is:

$$\tan(\theta) = v_z / \text{sqrt} (v_x^2 + v_y^2)$$

$$\tan(\theta) = 6 / \text{sqrt} (2^2 + (-2)^2) \approx 2.1$$

$$\theta \approx \tan^{-1}(2.1) \approx 64.7°$$

Sometimes we have to go the opposite way. Instead of being given the force vector, we know the magnitude of the force and the angle it makes with the horizontal and want to derive the corresponding vector from that. We will limit ourselves to two dimensions now as three dimensions would require an additional angle, for example the angle the vector makes with the y axis, and the math would become more involved. So two dimensions it is. Given the magnitude F of a force acting in the x-z-plane and the angle θ it makes with the horizontal, we can use the sine and cosine function to project the force onto the horizontal and vertical axis:

$$F_x = F \cdot \cos(\theta)$$

$$F_z = F \cdot \sin(\theta)$$

The resulting vector is $\mathbf{F} = (F_x, F_z)$. For example, a force of 50 N making an angle of 20° with the horizontal can be conveniently expressed by the vector $\mathbf{F} = (47.0 \text{ N}, 17.1 \text{ N})$ since $F_x = 50 \cdot \cos(20°) \approx 47.0$ and $F_z = 50 \cdot \sin(20) \approx 17.1$. With these basic vector techniques in mind, you can already solve a vast amount of pretty sophisticated problems involving forces.

Example 28:

A plane takes off in direction of the x-axis. The difference of thrust and frictional force gives rise to the horizontal force F_1 = (6,500 N, 0 N, 0 N). In addition to that, the plane experiences the lift F_2 = (-2,800 N, 0 N, 3,500 N). Calculate the net force on the plane, the corresponding magnitude and the angle the net force makes with the horizontal.

Solution:

We add the forces acting on the plane component by component to compute the net force:

$F = F_1 + F_2$

F = (6,500 N, 0 N, 0 N)

+ (-2,800 N, 0 N, 3,500 N)

F = (6,500 - 2,800 N, 0 N, 3,500 N)

F = (3,700 N, 0 N, 3,500 N)

The magnitude of the net force is:

$F = sqrt (3,700^2 + 0^2 + 3,500^2)$ N

$F \approx 5090$ N

And the angle to the horizontal:

$tan(\theta) = F_z / sqrt (F_x^2 + F_y^2)$

$tan(\theta) = 3,500 / sqrt (3,700^2 + 0^2)$

$tan(\theta) \approx 0.95$

$\theta \approx tan^{-1}(0.95) \approx 43.4°$

Example 29:

A motor drives a boat of mass m = 150 kg with a force F_1 = (600 N, 200 N). Besides the drag force F_2 = (-400 N, -150 N), the boat also experiences a force F_3 = (-170 N, 0 N) from a local ocean current. Note that since the motion takes place on a water surface, we can do without the z-component.

a) Compute the net force vector and its magnitude.

b) Compute the acceleration vector and its magnitude.

Solution:

a)

The net force is the vector sum of the forces involved. So simply add the vectors component by component.

$F = F_1 + F_2 + F_3$

$F = (600\ N,\ 200\ N)$

$+ (-400\ N,\ -150\ N)$

$+ (-170\ N,\ 0\ N)$

$F = (600 - 400 - 170\ N,\ 200 - 150\ N)$

$F = (30 N, 50 N)$

The net force accelerates the boat with 30 N in x-direction and 50 N in y-direction. The magnitude of the net force is (remember to sum the squares and apply the square root):

$F = sqrt (30^2 + 50^2) N \approx 58.3 N$

b)

We can calculate the acceleration from the vector form of Newton's second law. This means dividing each component individually by the total mass.

$a = F / m$

$a = (F_x / m, F_y / m)$

$a = (30 N / 150 kg, 50 N / 150 kg)$

$a = (0.2 m/s^2, 0.33 m/s^2)$

So the boat accelerates with 0.2 m/s² in x-direction and 0.33 m/s² in y-direction. The overall acceleration is:

$a = sqrt (0.2^2 + 0.33^2) N \approx 0.39 m/s^2$

Example 30:

Two people push a box along the x-axis. Person 1 applies a force $F_1 = 40 N$ at an angle of $\theta_1 = 25°$ to the horizontal and person 2 a force $F_2 = 60 N$ at an angle $\theta_2 = 35°$.

a) Convert both forces into vectors.

b) Compute the net force on the box.

Solution:

a)

We can use the sine and cosine function to project the force onto the horizontal and vertical axis. Using the corresponding formulas for person 1 leads to:

$F_x = 40 \, N \cdot cos(25°) \approx 36.3 \, N$

$F_z = 40 \, N \cdot sin(25°) \approx 16.9 \, N$

In summary:

$\mathbf{F_1} = (36.3 \, N, \, 16.9 \, N)$

For person 2 we get:

$F_x = 60 \, N \cdot cos(35°) \approx 49.1 \, N$

$F_z = 60 \, N \cdot sin(35°) \approx 34.4 \, N$

Which leads to the vector:

$\mathbf{F_2} = (49.1 \, N, \, 34.4 \, N)$

b)

We sum the forces to find the net force:

$\mathbf{F} = \mathbf{F_1} + \mathbf{F_2}$

$\mathbf{F} = (36.3 \, N, \, 16.9 \, N)$

114

+ *(49.1 N, 34.4 N)*

F = *(36.3 + 49.1 N, 16.9 + 34.4 N)*

F = *(85.4 N, 51.3 N)*

So in total the box experiences a push of 85.4 N in x-direction and 51.3 N in z-direction. By the way: is the 51.3 N upward push wasted energy when moving the box? Not at all. The upward force counters gravity and thus reduces the normal force between the ground and the box. The consequence: less ground friction.

Appendix

Unit Prefixes

peta (P) = 1,000,000,000,000,000 = 10^{15}

tera (T) = 1,000,000,000,000 = 10^{12}

giga (G) = 1,000,000,000 = 10^9

mega (M) = 1,000,000 = 10^6

kilo (k) = 1,000 = 10^3

deci (d) = 0.1 = 10^{-1}

centi (c) = 0.01 = 10^{-2}

milli (m) = 0.001 = 10^{-3}

micro (μ) = 0.000,001 = 10^{-6}

nano (n) = 0.000,000,001 = 10^{-9}

pico (p) = 0.000,000,000,001 = 10^{-12}

femto (f) = 0.000,000,000,000,001 = 10^{-15}

Unit Conversions

Since we often need to convert units from the United States customary system (USCS) to the metric (SI) system and vice versa, here's a list of commonly needed conversion factors.

Lengths, SI to USCS:

- Multiply meters with 3.28 to get to feet

- Multiply meters with 1.09 to get to yards

- Multiply meters with 0.00062 to get to miles

- Multiply kilometers with 3281 to get to feet

- Multiply kilometers with 1094 to get to yards

- Multiply kilometers with 0.62 to get to miles

Lengths, USCS to SI:

- Multiply feet with 0.30 to get to meters

- Multiply feet with 0.00030 to get to kilometers

- Multiply yards with 0.91 to get to meters

- Multiply yards with 0.00091 to get to kilometers

- Multiply miles with 1609 to get to meters

- Multiply miles with 1.61 to get to kilometers

To convert a squared to a squared unit, use the square of the conversion factor. For example you multiply m² by $3.3^2 \approx$ 10.9 to get to ft². In a similar fashion, using the cube of the conversion factor, you can convert cubed units.

Speeds:

- Multiply m/s with 3.6 to get to km/h

- Multiply m/s with 2.23 to get to mph

- Multiply km/h with 0.28 to get to m/s

- Multiply km/h with 0.62 to get to mph

- Multiply mph with 0.45 to get to m/s

- Multiply mph with 1.61 to get to km/h

Other commonly used units:

- Multiply pounds with 0.45 to get to kilograms

- Multiply kilograms with 2.22 to get to pounds

- 1 liter = 0.001 m³

- Multiply liters with 0.62 to get to gallons

- Multiply gallons with 3.79 to get to liters

- Celsius to Fahrenheit: °F = 1.8 · °C + 32

- Fahrenheit to Celsius: °C = 5/9 · (°F - 32)

- Kelvin to Celsius: °C = K - 273.15

- Celsius to Kelvin: K = °C + 273.15

Excerpt

*As a thank you to those have read this far, and hopefully enjoyed the journey, here's a small excerpt from the book **Physics! In Quantities and Examples**, an informal introduction to the basics of physics for beginners by yours truly Metin Bektas.*

To your skin, it makes a big difference if a certain amount of force is applied via the palm of a hand or the tip of a needle. Sometimes force F (in N) is just part of the picture. For many applications we also need to know the area A (in m^2) over which it is distributed. It can make all the difference between comfortable and painful, between enduring and breaking. The quantity pressure p (in N/m^2 = Pascal = Pa) reflects this. It specifies the amount of force applied per unit area:

$p = F / A$

For example, if we apply a force of F = 10 N using the palm of a hand (A = 0.01 m^2), then the pressure is p = 1000 Pa. While this sounds like a lot, it is much less than the air pressure at sea level. Pascal is a "small" unit, one Pascal is equivalent to a tree leaf on your shoulder, hardly noticeable. The area of a needle tip is around A = 10^{-7} m^2, so the same force produces a pressure of p = 10^8 Pa. This sounds like a lot and it really is. The pressure at the deepest point of the ocean, the Mariana Trench, about 11 km below the surface, is comparable to that.

Because Pascal is an impractical unit for everyday use, physicist often prefer alternatives such as the standard atmosphere (atm). 1 atm is equivalent to the mean air pressure at sea sea level, 101,325 Pa. The commonly used unit bar is roughly equivalent to the standard atmosphere. 1 bar is defined as 100,000 Pa. Another unit of pressure you might come across is pound per square inch (psi). You multiply it by 6895 to get to the SI unit Pascal.

From the Smallest to the Largest

Again we take a trip through the universe to get familiar with typical values of pressure. The lowest pressure ever measured is 10^{-17} Pa. It is the pressure you would experience when traveling in intergalactic voids. Within the Milky Way galaxy, in the void between the stars, the pressure is about one-hundred times greater than that, 10^{-15} Pa. The lowest pressure achieved in a laboratory is 10^{-12} Pa, which is below the atmospheric pressure on the moon (10^{-11} Pa).

Over the pressure in low Earth orbit (10^{-8} Pa) and the radiation pressure of sunlight ($4.7 \cdot 10^{-6}$ Pa, as calculated in the previous section), we get to the regions that humans are able to perceive. The pressure of a dollar bill is around 1 Pa, a quarter is a heavyweight compared to that (90 Pa). The overpressure necessary to break a window is around 5000 Pa, the pressure inside a vacuum cleaner tops that by a factor of sixteen (80,000 Pa). No wonder cats are so scared of them.

As stated, the standard atmospheric pressure at sea level is 101,325 Pa = 1 atm. On Venus the atmosphere is much thicker (92 atm). This combined with the 400 °C temperature and presence of sulfuric acid would kill a person within seconds. It is still nothing compared to the 1100 atm pressure at the bottom of the Mariana Trench. And hi-tech water jet cutters can reach up to six times that (6000 atm). A few scientists obviously thought they can do better and produced a water jet with a pressure of 690,000 atm, enough to synthesize diamonds.

Again, nature easily tops all of this. The pressure inside Earth's core is estimated to be around 3.6 million atm, the sun even goes up to 250 billion atm. This would crush a person within seconds into the dot you see at the end of this sentence. As usual, the most extreme values in the observable universe can be found in neutron stars: 10^{29} atm, that's a one with twenty-nine zeros. Anything beyond that would result in a black hole.

Hydraulic Press

A hydraulic press is a system using fluids to amplify forces. It allows the lifting of heavy objects with a relatively small force input. Pascal's principle makes this possible. It states that pressure is transmitted undiminished throughout an enclosed fluid. In other words: if you increase pressure at one point of the confined fluid, the pressure will increase by the same amount at any other point. So pressure "travels" without loss through the fluid.

A result from Pascal's principle is that in the absence of gravitation, the pressure at any point in an enclosed fluid would be the same. Gravitation complicates matters a bit by adding weight, meaning that the fluid at the top is under less pressure than the fluid at the bottom. However, all points at the top still experience the same pressure.

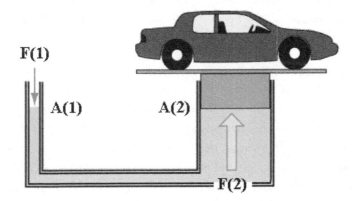

With this out of the way, we can take a look at the hydraulic press. It consists of an enclosed fluid with two moveable cylinders. We'll denote the area of the cylinders by A(1) and A(2). Since both are at the top of the fluid, the cylinders experience the same pressure.

p(1) = p(2)

And since pressure is computed by dividing force by area, we can write the identity as such:

F(1) / A(1) = F(2) / A(2)

with F(1) and F(2) being the forces acting on the respective cylinders. This equation is where the magic is. Suppose for example the cylinder areas are A(1) = 1 m² and A(2) = 12 m². We push cylinder 1 downwards with a force of F(1) = 10 N. How does this affect cylinder 2? Inserting the given values into the equation, we get:

10 N / 1 m² = F(2) / 12 m²

F(2) = 120 N

So we amplified the input force by a factor of twelve, enabling us to lift objects that would normally be too heavy. You might be thinking to yourself: What sorcery is this? Isn't this a violation of energy conservation? What's the catch? Indeed, there is a catch. If we amplify the force by a factor of 12, the same happens to distance. So to lift cylinder 2 by 1 m, we would need to move cylinder 1 by 12 m. This means that the conservation of energy (force times distance) is still valid. We can't gain energy with the hydraulic press, we gain force at the cost of distance.

Air Pressure

The air pressure at a certain point is determined by the weight of the air per unit area above said point. At sea level and under normal circumstances, a 1 m² column of air extending all the way into space weighs 101,325 N (which corresponds to a mass of 10,330 kg), hence the pressure p = 101,325 Pa. If meteorological events cause the pressure to drop, it just means that now there's less air above you.

The lowest pressure ever recorded at sea level was 87,000 Pa. It was measured 1979 in the Pacific Ocean during the storm "Tip". This value leads us to conclude that the weight of the 1 m² air column dropped to 87,000 N (or 8870 kg). So the mass of air resting on the ocean surface was approximately 1500 kg less than usual.

The above definition of air pressure explains why at higher altitudes we can expect lower pressures. As you go up, the amount of air above you becomes smaller. The barometric formula provides a good estimate for the air pressure p (in Pa) at a certain height h (in m) above sea level:

$p = 101,325 \cdot exp(-0.00012 \cdot h)$

According to the National Geographic Magazine, the highest city in the world is La Rinconada in Peru. It has 50,000 inhabitants and is located at a height of 5100 m above. The growth of the city is driven by gold. Most of the people who come to La Rinconada hope to make a fortune in the nearby gold mine and are willing to endure cold weather and the thin air, that makes any hard labor even worse. The air pressure is only around 55,000 Pa. This means that in terms of air mass, you are already about halfway through Earth's atmosphere in La Rinconada.

(Picture of La Rinconada, Peru, a city halfway through Earth's atmosphere)

Magdeburg Hemispheres

Otto von Guericke, a 17th century German scientist, thought up an interesting experiment demonstrating the power of air pressure. It was first performed in 1654 for the Emperor Friedrich III in Regensburg and soon became a popular experiment for entertainment purposes and physics lectures. For von Guericke, the experiment provided a great way to demonstrate his revolutionary invention: the world's first vacuum pump.

In the experiment, von Guericke used two identical copper hemispheres of radius $r = 25$ cm that were put together to a sphere and sealed. The vacuum pump was connected and the air pumped out, creating a near vacuum inside the sphere. Then thirty horses, in two teams of fifteen, tried to pull the *hemispheres apart - without success. Only after von Guericke opened a valve to equalize the pressure could the hemispheres, which became known as the Magdeburg hemispheres, be separated.*

Let's take a look at this demonstration from a quantitative point of view. How much force is necessary to separate the hemispheres? The basis of the calculation is the equation for pressure, solved for the force F:

$F = p \cdot A$

For a perfect vacuum inside the sphere, we can set p = 101,325 Pa, the air pressure at sea level. A detailed mathematical analysis shows that rather than using the surface area of the sphere, we have to input the area bound by the joint of the hemispheres (a circle of radius 25 cm):

$A = \pi \cdot (0.25 \ m)^2 \approx 0.1963 \ m^2$

Thus the force necessary to separate the hemispheres is:

$F \approx 101,325 \ Pa \cdot 0.1963 \ m^2 \approx 19,900 \ N$

This force corresponds to the weight of a 2000 kg object - no wonder the horses tried in vain. The experiment is still popular today, you might come across it someday, minus the horses, in a university or museum. In 2002, on the occasion of von Guericke's 400th birthday, several sculptures of the Magdeburg hemispheres were put up in Magdeburg. They serve as a reminder that science does not need to be abstract and great ideas don't need to be complex. The power of this demonstration lies with its simplicity. It provides a deep insight into the world around us without relying on prior knowledge.

Copyright and Disclaimer

Made in the USA
Coppell, TX
11 April 2021